Volumetric and Phase Behavior of Oil Field Hydrocarbon Systems

M. B. Standing

Society of Petroleum Engineers of AIME

Dallas 1977

Ninth Printing
1981

Library of Congress Catalog Card No. 52-11180

Printed in the United States of America by
Millet the Printer, Inc.
Dallas, Texas

PREFACE

During the past twenty years petroleum engineers have developed many specialized methods of calculating the pressure-production characteristics of oil and gas reservoirs. Part of the basic information required in these calculations pertains to the physical properties of the reservoir fluids and to the manner in which the properties change as the reservoir pressure changes. Naturally, many relevant papers have appeared in the technical literature. Some have pointed out the manner by which the fluid properties have been measured in the laboratory. Others have pointed out the manner by which the fluid property data have been used to calculate reservoir behavior. Yet, only a few authors have coordinated the two and, what is most important, have shown how the measured laboratory data can be altered to fit specific reservoir conditions.

The types of hydrocarbon systems encountered in oil field operations range from almost pure methane gas to heavy bituminous material having the nature of road tar or asphalt. In an effort to systematize the discussion of the phase behavior of these complex systems, they have been divided into three categories: (a) gases, (b) condensate systems, and (c) dissolved gas systems. Just where the dividing line between gases and condensate systems lies is difficult to say, although condensate systems and dissolved gas systems can be differentiated easily by the critical temperature of the system. Supporting framework to the discussion of these three systems is given in the chapters on general phase behavior of hydrocarbon systems, sampling of oil reservoirs, and material balance calculations.

Conflict in the nomenclature of fluid flow and thermodynamics was encountered early in the preparation of the book. As a result, it was necessary to readjust some of the symbols from their usual meaning in order not to have duplication. A complete list of the nomenclature and definition of terms used in this book follows Chapter 8.

The author has attempted to present as many charts as possible to assist in reservoir engineering calculations. In this connection, the Appendix contains three calculating charts that are helpful in estimating properties of naturally occurring hydrocarbon systems at reservoir conditions from the field parameters of gas-oil ratio, gas gravity, and oil gravity. While the data from which the charts were constructed are exclusively "California data," there are no obvious reasons why the charts will not apply to systems from other locations.

The author wishes to acknowledge the kind interest of the Management of the California Research Corporation and others during the processing of this book. Mr. A. L. Vitter, Jr., of The California Company and Messrs. R. L. Parsons and R. E. Loeck of California Research Corporation acted as a reviewing committee during the preparation of the book and made valuable suggestions toward the final form. Mrs. B. Hansen, Mrs. V. Cox, and Mrs. E. K. LaRue were particularly helpful in preparing charts and in proofreading.

M. B. STANDING

July, 1952
La Habra, California

Contents

*Charts will be found in the back of the book.

CHAPTER 1

INTRODUCTORY PHASE BEHAVIOR AND FLUID FLOW CONCEPTS

Hydrocarbon systems found in petroleum reservoirs are known to exhibit multiphase behavior at temperatures as low as −250°F and as high as 1000°F. Many systems exist as two phases at pressures as high as 15,000 psia. Because the molecular composition of the gas and liquid phases in equilibrium with each other in a reservoir rock is different, the physical properties of the phases differ. This difference in phase properties, coupled with the ability of reservoir rock to transmit gas and liquid with different degrees of ease, causes the complex behavior of oil reservoirs.

A few of the basic concepts of phase behavior and fluid flow in porous media will be presented in this chapter. The phase behavior discussions will start with the behavior of a pure compound and will proceed from there to discussions of binary and multicomponent system behavior. Following the phase behavior discussions, basic concepts of flow of multiphase fluids through porous media will be outlined. Correlating the ideas regarding phase behavior of multicomponent systems with expected flow behavior in porous reservoir rock gives a basis for the discussions in subsequent chapters.

Pure Compounds (Ethane)

A portion of the pressure-temperature diagram of ethane is shown in Figure 1. This diagram is used principally to determine whether ethane exists as a liquid, a vapor, or a combination of both, at a specified pressure and temperature. Liquid and vapor exist in equilibrium at all points on the curve; only one phase can exist at pressures and temperatures removed from the curve. The vapor pressure curve terminates at the critical temperature and critical pressure, C. For pure compounds, the critical temperature can be defined as the temperature above which two phases cannot exist, regardless of the pressure upon the system. Likewise, the critical pressure is that pressure above which vapor and liquid cannot exist in equilibrium.

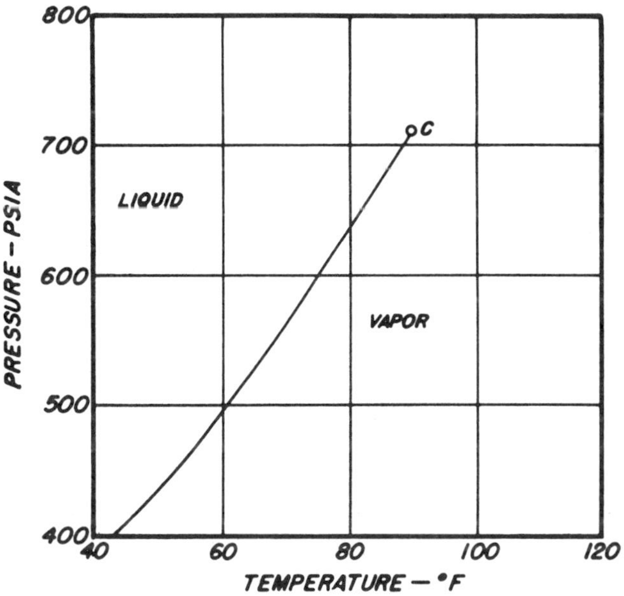

Figure 1—Pressure-Temperature Diagram of Ethane

The pressure-volume relation of ethane is shown in Figure 2. The area within the dashed line represents conditions under which two phases can exist in equilibrium; the area to the right represents vapor, while the area to the left represents liquid. Point C again designates the critical pressure and temperature (709 psia and 90°F, respectively). The line AC defines what are generally called bubble point or saturated liquid conditions, and line CB designates the properties of ethane at dew point or saturated vapor conditions.

To illustrate the use of Figures 1 and 2, let us assume that one pound of ethane is contained in 0.25 cu ft of space at 60°F. At the specified temperature and volume conditions, ethane will be in vapor form and exert a pressure of 465 psia. This is shown in Figure 2. When the ethane vapor is compressed at a constant temperature of 60°F, liquid will condense first at 495 psia. Point *B*, the state at which the system is entirely in the vapor form except for an infinitesimal amount of equilibrium liquid, is, by definition, the dew point. Further

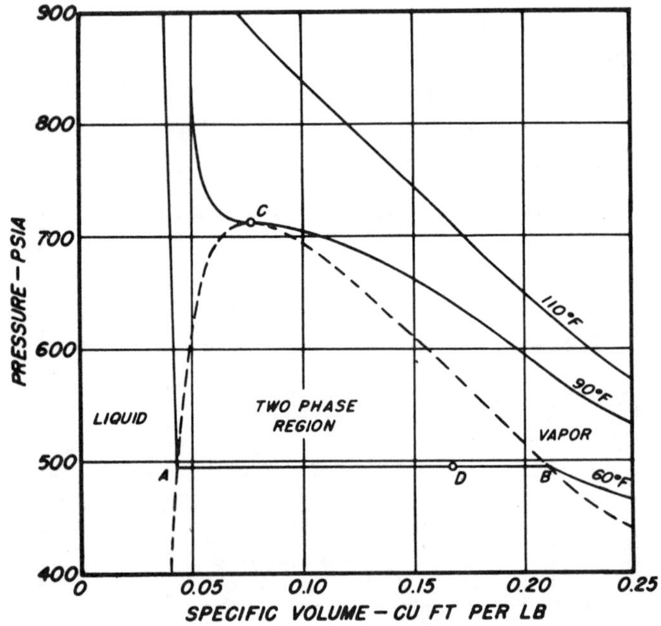

Figure 2—Pressure-Volume Diagram of Ethane

decreases in the volume of the system will result in the formation of larger and larger proportions of liquid in the system, until at point *A* all material will be in the liquid condition, except for an infinitesimal amount of equilibrium vapor (bubble point). Additional decrease in volume will result in rapid increase in pressure because of the low compressibility of liquid ethane. During the process of moving between points *A* and *B*, the specific volumes of the two phases will remain constant at 0.0423 cu ft per lb of liquid and 0.2105 cu ft per lb of vapor. A simple weight balance shows that when the system as a whole has an intermediate specific volume, represented by point *D*, the ratio of weight of liquid to vapor will be in proportion to the length *BD* to the length *DA*.

The behavior of other pure compounds is similar to that outlined for ethane. A vapor pressure chart for the lighter hydrocarbons is shown in Figure 3. Table 1 lists important physical properties of the light paraffin hydrocarbons, as well as those of several miscellaneous compounds.

Binary Systems

The phase behavior of two-component systems is slightly more complex than that of pure compounds. The main differences are: (1) the pressure-temperature relationships can no longer be represented by a vapor pressure curve, and (2) a different definition of critical temperature and pressure must be used to describe the phase behavior.

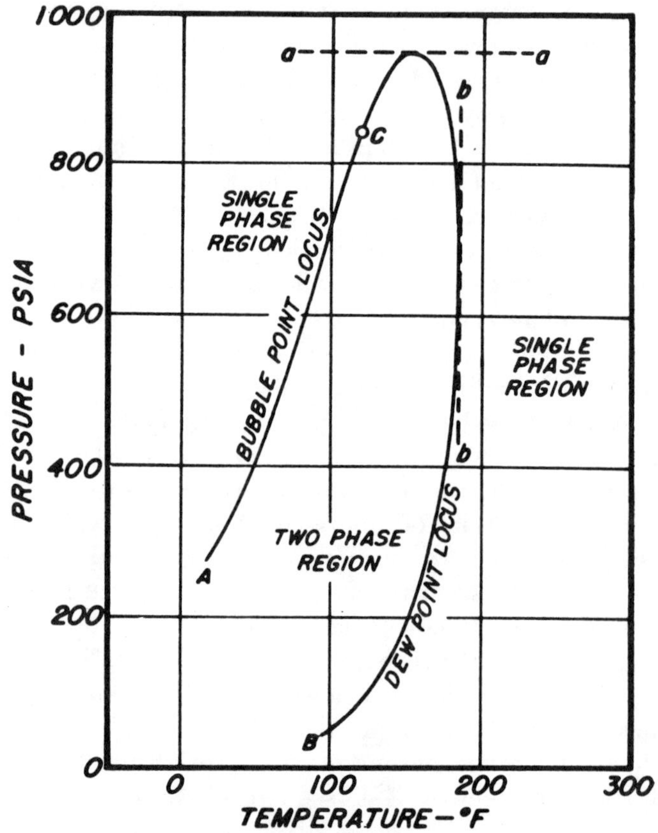

Figure 4—Pressure-Temperature Diagram of the Ethane-*n*-Heptane Mixture Containing 96.83 Mole Per Cent Ethane[1]

A portion of the phase behavior of a mixture containing 96.83 mole per cent ethane (90.22 weight per cent) and 3.17 mole per cent *n*-heptane is shown in Figure 4. As with all binary systems, the pressure-

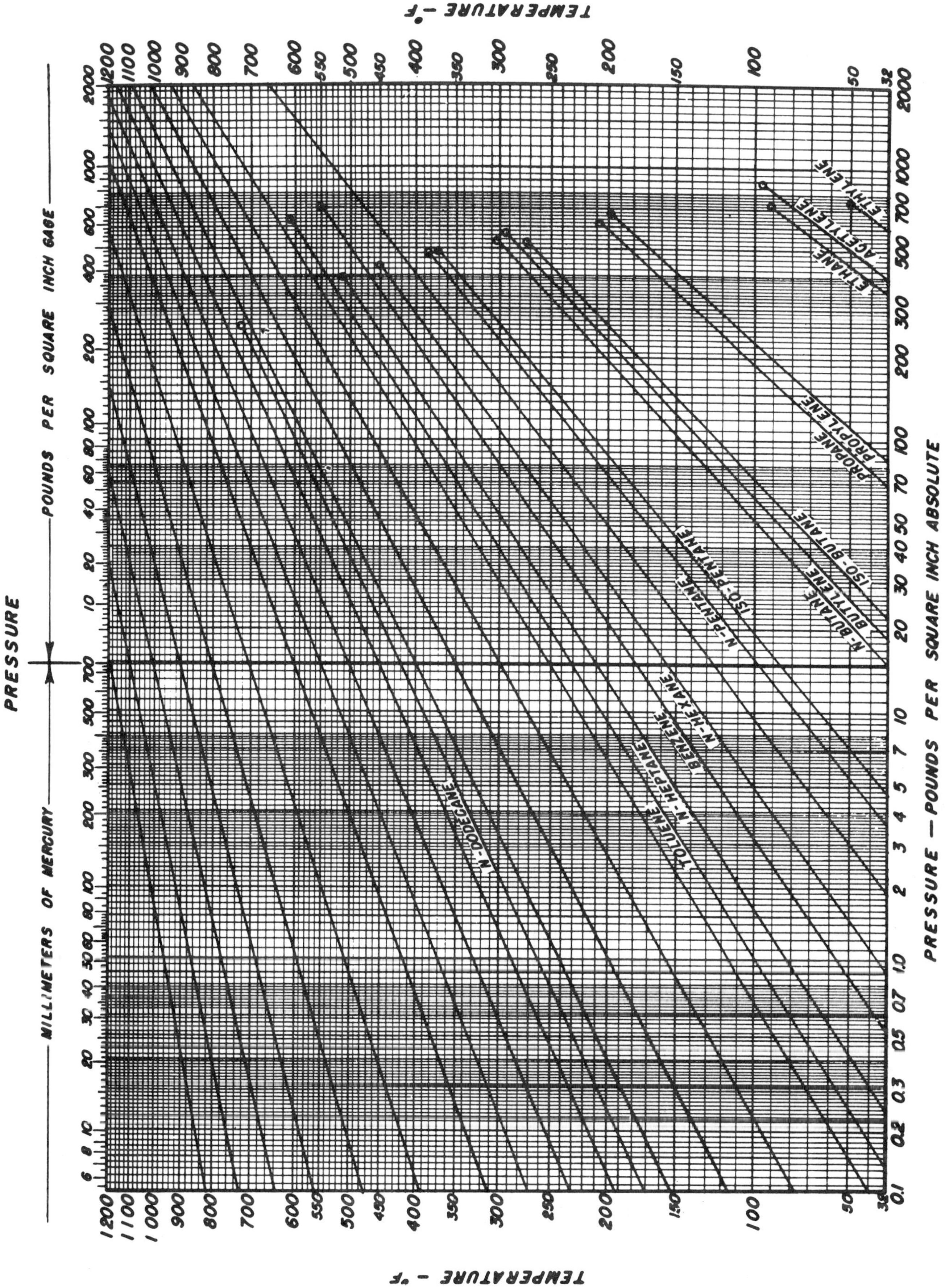

Figure 3—Vapor Pressure Chart for Hydrocarbons

TABLE I

PHYSICAL PROPERTIES OF THE PARAFFIN HYDROCARBONS & MISCELLANEOUS COMPOUNDS

COMPOUND	MOLECULAR WEIGHT	BOILING POINT AT 14.7 PSIA °F	CRITICAL CONSTANTS		LIQUID DENSITY 60°F, 14.7 PSIA		GAS DENSITY 60°F, 14.7 PSIA (PERFECT GAS)		CU FT GAS PER GALLON LIQUID (ACTUAL)
			PRESSURE PSIA	TEMPERATURE °R	Gm. PER cc	POUNDS PER CU FT	POUNDS PER MCF	CU FT GAS PER GALLON LIQUID	
METHANE	16.04	−258.7	673	344	—	—	42.27	—	—
ETHANE	30.07	−127.5	709	550	—	—	79.23	—	—
PROPANE	44.09	−43.7	618	666	0.5072	31.66	116.19	36.35	35.78
ISO-BUTANE	58.12	10.9	530	733	0.5625	35.12	153,15	30.59	29.70
N-BUTANE	58.12	31.1	551	766	0.5836	36.43	153.15	31.75	30.77
ISO-PENTANE	72.15	82.1	482	830	0.6241	38.96	190.11	27.35	26.17
N-PENTANE	72.15	96.9	485	847	0.6305	39.36	190.11	27.63	26.36
N-HEXANE	86.17	155.7	434	915	0.6637	41.43	227.07	24.34	22.83
N-HEPTANE	100.2	209.2	397	973	0.6875	42.92	264.03	21.69	19.95
N-OCTANE	114.2	258.2	370	1025	0.7062	44.09	300.99	19.55	17.77
N-NONANE	128.3	303.4	335	1073	0.7211	45.02	337.95	17.78	15.88
N-DECANE	142.3	345.2	312	1115	0.7333	44.78	374.91	16.30	14.30
AIR	28.97	−317.7	547	239					
NITROGEN	28.02	−320.4	492	227	—	—			
OXYGEN	32.00	−297.4	732	278	—	—			
CARBON DIOXIDE	44.01	−109.3	1072	548	—	—			
HYDROGEN SULPHIDE	34.08	−76.5	1306	673					
WATER	18.02	212.0	3206	1165	0.9991	62.37			

temperature diagram takes the form of an envelope rather than the single vapor pressure curve of pure compounds. The area within the envelope defines the temperature and pressure range wherein vapor and liquid can exist in equilibrium. Outside the envelope only one phase can exist. The dashed line *aa* is the maximum pressure at which two phases can exist. The line *bb*, defining the maximum temperature at which two phases can exist in equilibrium, is called the "cricondentherm."

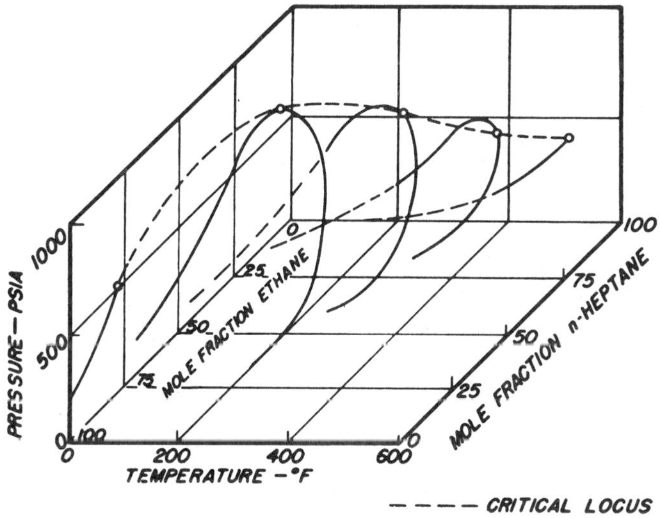

Figure 5—Pressure-Temperature-Composition Diagram of the Ethane-*n*-Heptane System

Point *C* in Figure 4 is the critical pressure and temperature of this particular mixture. It is apparent from inspection of Figure 4 that the definition of the critical temperature and pressure used for a pure compound cannot apply to the binary mixture, as multiple phases exist at pressures and temperatures greater than the critical values. A better general definition of the critical state is that state at which **all** intensive* properties of the liquid and vapor phase become equal. This definition applies equally well to pure compounds, but is usually neglected in favor of the simpler definition.

The critical pressure, critical temperature, cricondentherm, and maximum pressure value are functions of the composition of the particular mixture. For example, in the ethane-*n*-heptane system the variables of pressure, temperature, and composition are as shown in Figure 5. In Figure 4, it will be noted that the critical state of the 96.83 mole per cent ethane mixture is well to the left and below the maximum two-phase pressure. In Figure 5, it can be seen that as the per cent of *n*-heptane in the mixture increases, the critical state moves toward the right side of the phase envelope, until at 75 mole per cent *n*-heptane it lies well to the right and below the maximum pressure.

* Intensive properties are those which are independent of the amount of material; for example, density, pressure, temperature, viscosity, chemical potential, and surface tension. Extensive properties depend on the amount of material; for example, area, weight, inertia, and volume. The product of an intensive and extensive property is a measure of energy; for example, pressure × volume = mechanical energy.

Multicomponent Systems

The phase behavior of multicomponent systems is much the same as that of the ethane-*n*-heptane system. As will be shown later, knowledge of the critical temperature and pressure of multicomponent systems is useful in describing and correlating the phase behavior of such systems. As a matter of fact, from the critical temperature alone, one can make reasonable estimates regarding PVT behavior of any reservoir mixture.

As an example of how knowledge of the critical state can be used, let us assume the particular mixture of natural gas and gasoline studied by Katz and Kurata[2] to be representative of the hydrocarbon system existing in a particular reservoir. For the moment we shall not consider how we know it represents the reservoir material, but will assume that it does, and that we have a small quantity of the mixture isolated in a laboratory pressure cell in which we can observe the phase behavior at various pressures and temperatures. This behavior is shown in Figure 6.

The first thing to be noted is that the general shape of the two-phase envelope in Figure 6 is similar to the envelopes exhibited by binary systems. As in binary systems, the loci of bubble points and dew points meet and join as a smooth curve at the critical state. In fact, the only major difference between the natural gas-natural gasoline and ethane-*n*-heptane phase diagrams is the higher critical pressure exhibited by the multicomponent system.

In addition to assuming that the data shown in Figure 6 represent the phase behavior of the reservoir fluid, let us also assume original reservoir conditions of 3000 psia and 190°F. Referring to Figure 6, it would be expected that when the reservoir pressure had fallen to 2690 psia, a liquid phase would appear throughout the reservoir. The liquid would continue to increase to a maximum

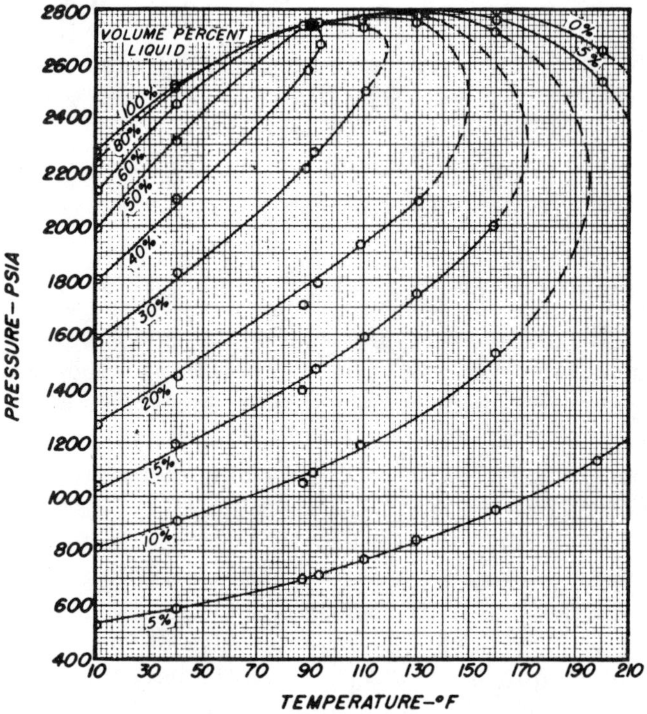

Figure 6—Phase Behavior of a Natural Gas-Natural Gasoline Mixture (After Katz and Kurata, Ind. Eng. Chem., 1940)

of 11 per cent at 2200 psia, after which further pressure reduction would result in the revaporization of the liquid phase.*

The formation of liquid from a vapor as a result of decrease in pressure is termed "retrograde condensation." The opposite process of vaporization of a liquid as a result of an increase in pressure is called "retrograde vaporization." There can exist, of course, the comparable isobaric processes of retrograde vaporization and condensation as, for example, at pressures above 2750 psia (the critical pressure) in Figure 6.

* The data of Figure 6 were determined for a condition of constant system composition; i.e., the reduction of pressure was obtained by increasing the volume of the container in which the system resided. This obviously does not conform to reservoir behavior wherein the pressure reduction is a result of removing material from the system (variable composition). However, the assumed process is satisfactory for the present discussion.

An easy way of thinking of retrograde behavior is to use the word retrograde as meaning "in opposite direction to that usually observed." Thus, the formation and growth of a liquid phase as a result of isobaric increase in temperature are opposite from the usual behavior, as pure compounds always vaporize (decrease liquid phase) upon isobaric increase in temperature. Likewise, our knowledge of the behavior of pure compounds indicates that an increase in pressure, at constant temperature, will result in condensation of vapor into liquid (Figure 2); therefore, the opposite behavior of binary and multicomponent systems causing a liquid phase to decrease in extent while undergoing an increase in pressure is called a retrograde phenomenon.

Referring again to Figure 6, it will be noted that had the reservoir temperature been 70°F rather than 190°F, a bubble point, rather than a dew point, would have been reached at 2690 psia. This would have resulted in the appearance of gas in a sand saturated with liquid, rather than the appearance of liquid in a sand saturated with vapor. Consequently, one would expect some difference in the pressure-production behavior of the reservoir and some difference in the manner by which the reservoir should be produced to obtain the maximum recovery of hydrocarbons.

Two important facts are to be learned from the above:

(1) The composition of a hydrocarbon system defines the critical pressure and temperature of the system. (See Figure 5.) As the proportion of light hydrocarbons (methane, ethane) in the system increases, the critical temperature of the system decreases.

(2) The relation of the critical temperature of the hydrocarbon system to the reservoir temperature will determine whether retrograde behavior will be encountered. Whether the system exhibits a dew point or a bubble point at reservoir temperature determines the flow mechanism in the reservoir sand.

Rigid application of these principles to oil reservoir systems is difficult, as few data are available regarding the critical temperature of crude oil-natural gas systems as a function of composition.

The fact that oil field usage recognizes only two components, oil and gas, each of which is a multicomponent system, makes generalizations difficult; nevertheless, the following qualitative statements can be made:

(1) Systems existing in one phase and having gas-oil ratios greater than 10,000 cu ft per bbl are likely to have high concentrations of methane and the other light hydrocarbon components. Therefore, the critical temperature of the systems will be low (perhaps of the order of that of methane, −116°F). Consequently, at reservoir temperatures of 150 to 300°F, one may expect that such a system will exhibit isothermal retrograde phenomena.

(2) Systems existing in one phase and having gas-oil ratios less than 2000 cu ft per bbl are likely to have critical temperatures higher than the reservoir temperature. These systems will have bubble points, and initial production will be by dissolved gas drive.

(3) Systems in the intermediate range of gas-oil ratio may exhibit either bubble point or dew point behavior. It is necessary to make laboratory tests to determine the actual behavior of the system at reservoir conditions.

Multiphase Fluid Flow

When pores in reservoir rocks contain more than one fluid, the permeability of the rock is different to each fluid. The permeability to water or gas that a given rock matrix will have is primarily a function of the water or gas saturation within the rock pores.* The permeability of the rock to oil is affected by the fraction of the pore occupied by oil (oil saturation), as well as by the relative proportions of water and gas present. When pressure gradients exist in the fluids, the rates of flow of the individual phases will be as follows:

$$q_g \propto \frac{k_g}{\mu_g} \cdot \left(\frac{dP}{dx}\right)_g$$
$$q_o \propto \frac{k_o}{\mu_o} \cdot \left(\frac{dP}{dx}\right)_o \qquad (1)$$
$$q_w \propto \frac{k_w}{\mu_w} \cdot \left(\frac{dP}{dx}\right)_w$$

* This statement assumes that water is the fluid wetting the rock material and that gas is the fluid residing near the center of the pore. Liquid oil would be the intermediate phase.

where q_g, q_o, q_w are volumetric rates of flow of gas, oil, and water at reservoir conditions; k_g, k_o, k_w are effective permeabilities to the three phases; μ_g, μ_o, μ_w are viscosities of the three phases at reservoir conditions; and $(dP/dx)_g$, $(dP/dx)_o$, $(dP/dx)_w$ are pressure gradients within the three phases.

A reasonable representation of the relative flow rates of gas phase and liquid oil phase can be made by neglecting the behavior of the water phase and assuming equal pressure gradients in the two hydrocarbon phases. Dividing the first of Equation (1) by the second, and rearranging gives,

$$\frac{q_g}{q_o} \propto \frac{k_g}{k_o} \cdot \frac{\mu_o}{\mu_g} \qquad (2)$$

Equation (2) expresses the relative flow rates of gas and liquid oil phases under reservoir conditions of pressure, temperature, and fluid saturation. Two generalizations can be made, even though phase properties and reservoir rock properties vary between reservoirs and between areas within a single reservoir.

(1) The viscosity ratio, μ_o/μ_g, is generally greater than 10 and can be as high as 10^6. This is because reservoir gas viscosities are of the order of 0.01 to 0.07 centipoise[3] and oil phase viscosities vary from 0.2 to 10,000 centipoises.[4]

(2) The permeability ratio, k_g/k_o, may vary from zero to infinity because, over particular ranges of gas and liquid oil saturation, the permeability to either phase may be zero.

The gas and oil permeability curves of a Gulf Coast sandstone shown in Figure 7 are typical of data that have been determined for a large number of sandstones. The curves indicate the general ranges of fluid saturation wherein the permeability to one or the other of the hydrocarbon phases can be zero.

Let us speculate on the probable phase and fluid flow behavior in the region of a well, using in Equation (2) the information given by Figure 7. Let us consider also the phase behavior of the original reservoir fluid to be represented by the data of Figure 6, first taking the case of the reservoir fluid originally existing at 2800 psia and 70°F, and second, the fluid originally existing at 2800 psia and 190°F.

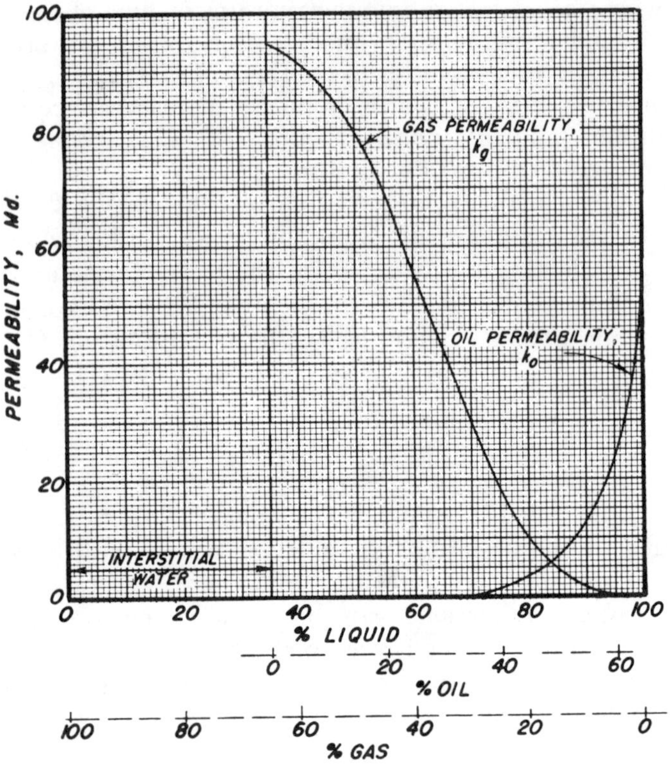

Figure 7—Multiphase Permeability Curves for a Consolidated Sandstone

The probable phase and fluid flow behavior of the fluid originally at 2800 psia and 70°F can be reasoned as follows:

(1) The original fluid exists as one phase at 2800 psia and 70°F, to which the reservoir rock has a 50 millidarcy permeability.

(2) As long as the reservoir pressure remains above the bubble point pressure of the fluid (2690 psia at 70°F), gas will not be formed in the reservoir rock pores. The formation of a gas saturation within the pores will start when the pressure has fallen below 2690 psia. As the pressure at the well bore will always be the lowest of the pressures causing flow to the well, gas will appear first in the rock pores adjacent to the well bore.

(3) For a while the gas coming out of solution in the liquid at pressure less than 2690 psia will have a tendency to remain in position in the rock pores. Figure 7 indicates the permeability to gas to be zero, until a gas saturation of at least 5 per cent of the pore volume is established.

(4) During the time the gas saturation is increasing in the region of the well bore, there will be some impairment of the well's capacity to produce oil because of the decrease in liquid phase permeability.

(5) The permeability ratio, k_g/k_o, increases rapidly after the critical gas saturation is established, and, in conjunction with the viscosity ratio, μ_o/μ_g, ultimately causes a high ratio of gas to liquid phase flow. When the gas saturation in the rock pores reaches 30 to 35 per cent, the permeability to oil may be so low as to cause abandonment of the well as an oil producer.

The above described process, whereby liquid oil is forced into the well bore by the generation and expansion of a gas phase, is termed a "dissolved gas drive." The possible ultimate yield of oil by this mechanism alone is about 25 per cent of the original oil in place. Additional yields are possible if the characteristics of the reservoir and contained fluids are such as to favor the functioning of additional mechanisms, such as gravity drainage and water drive.

A reasoning process similar to the above can be used to picture the phase and fluid flow of systems having critical temperatures lower than that at which they exist in reservoirs. The phase behavior of such a condensate system has already been discussed for the case of the 190°F isotherm in Figure 6. The permeability data of Figure 7 can be considered to be applicable to this type of system, as well as the dissolved gas system. Without going into the detail of the dissolved gas drive discussions, several facts are readily apparent from inspection of Figures 6 and 7.

(1) The maximum amount of condensate that would be formed by pressure reduction at 190°F is indicated in Figure 6 to be 11 per cent of the hydrocarbon system volume. Because of the 35 per cent interstitial water present in the rock pores, this 11 per cent would be equivalent to $11(1 - 0.35) = 7.1$ per cent of the rock pore volume.

(2) The data of Figure 7 indicate that at least 33 per cent condensate saturation is needed to have a liquid phase permeability greater than

zero. Thus, the 7.1 per cent condensate formed by retrograde action will be an immobile phase and will be lost insofar as flow into the well is concerned. However, this statement must be qualified as covering the behavior occurring in the major portion of the reservoir; the effect of converging flow lines in the area adjacent to the well will make it possible for some condensate to flow into the well bore.

Flash and Differential Equilibrium Processes

In the discussions to this point, two concepts have been implied as to the type of equilibrium process that occurs between hydrocarbon phases within the reservoir matrix. The data shown in Figure 6, which have been used to picture the phase behavior of fluids within reservoirs, were determined for a system of constant composition. In contrast to the constant system composition viewpoint, Equation (2) and the data of Figure 7 indicate that in the great majority of cases the hydrocarbon liquid and gas phases in a reservoir move at different rates. Because the gas and liquid phases in a multicomponent system are always of different composition,* the preferential movement of any one phase across the boundary of the system (reservoir) always results in a change in over-all composition of the remaining material. The two processes have been termed the "flash" process and the "differential" process. A flash process is one in which the over-all composition of a system remains constant; a differential process is one in which the over-all composition of a system varies as the process proceeds. The difference between these processes can be made clear by an example of each.

In obtaining the data of Figure 6, Katz and Kurata made use of a variable volume PVT cell and measured the phase volumes at several indicated pressures and temperatures. Because the composition of the system was not changed during the tests, the results are referred to as flash equilibrium data. An example of a dynamic flash equilibrium process is the separation of gas and oil in a field separator or trap. Over relatively long time periods the composition of the fluid entering the trap can be considered to be constant. Consequently, the composition and proportions of gas and liquid phases in the trap at any specified time are functions of only pressure, temperature, and the over-all system composition.

An example of a differential process can be visualized by using Figure 6 as a reference. Suppose the system studied was originally at 2800 psia and 70°F. Reduction in pressure to 2690 psia would cause the system to be at its bubble point condition; further reduction to 2689 psia would result in equilibrium gas and liquid phases. If it were possible to remove the free gas phase from the cell without disturbing the liquid phase, a new system having a bubble point of 2689 psia would remain. Continued stepwise removal of small increments of equilibrium gas would eventually bring the cell to atmospheric pressure via a series of systems having different compositions and bubble point pressures.

To summarize the difference between flash and differential processes, it might be said that the latter requires a knowledge of the exact route by which the process has progressed to a certain point, whereas the history of past events is immaterial in a flash process. In a flash process it is sufficient to state the original system composition and the desired pressure and temperature conditions; in a differential process the manner by which the original system degenerates to the desired pressure and temperature also must be defined.

More will be said about flash and differential processes in subsequent chapters.

* Except at the critical state. Composition is one of the intensive factors in which the vapor and liquid are equal at the critical.

CHAPTER 2

SAMPLING METHODS AND APPARATUS

The object of sampling is to obtain oil and gas for laboratory studies of physical and chemical properties of fluids present in underground reservoirs. Whereas the desired type of result determines methods of performing the laboratory tests, the characteristics of the reservoir fluids and the mechanical equipment used in completing the well determine the method to be used in obtaining the samples. Needless to say, obtaining a proper sample is as important as subsequent laboratory tests; yet few engineers understand the advantages and limitations of the several methods that are commonly used in sampling.

Before discussing their merits and limitations it may be well to consider several general facts involved in getting samples of oil and gas that are representative of reservoir fluids. In the first place, there is no assurance that any sample obtained from one well is representative of the fluid throughout the reservoir. Theoretically, the effect of the earth's gravity field should cause differences in composition of oil lying at different elevations within a reservoir. Also, the reservoir fluid may vary in composition between locations having the same structural elevation because of movement of the rock strata comprising the reservoir during geologic time. Both kinds of compositional variation have been observed in oil fields.

The second factor involves the question: To what extent is the fluid in the tubing representative of the reservoir fluid in the region of the well being sampled? This factor is important because all the methods make use, either directly or indirectly, of fluids obtained from the well tubing. Two reasons why the fluid flowing in the tubing might not be representative of the reservoir fluid are:

(1) Completion of the well may result in simultaneous production of fluids into the tubing from two or more zones. Thus, the sample obtained may be an unknown mixture of two fluids and will not be representative of either.

(2) Where both liquid and gas phases exist in intimate contact in the same zone, the flow rates of the two phases into the well bore are almost always different. Consequently, it is fortuitous if the flow rates are such as to cause the tubing fluid to have the same composition as the reservoir fluid as a whole.

The third factor involves the accuracy of sampling the fluid flowing in the tubing. While the field engineer has little control over the first two factors, he can do something about the third.

The greatest use of PVT data lies in calculating reserves of oil and gas in place in the reservoir and the effect of field operational methods on the recovery of oil and gas. Both calculations make use of the change with pressure of the physical properties of the reservoir fluid. Because it is difficult to evaluate the over-all composition and physical properties of a reservoir fluid (except possibly in the case of an under-saturated fluid), most PVT data are designed purposely to give only the properties of the individual phases. In fact, many tests may not evaluate directly the properties of the individual phases believed to be in the reservoir, but give data from which the desired property can be obtained by extrapolation. In either event, the over-all properties of the reservoir fluid are calculated from the individual phase properties and an estimate of the relative phase volumes present in the reservoir rock. The best possible estimates of phase properties during the entire life of a reservoir calls for the testing of more than one sample. This is brought out by the following three points:

(1) A sufficient number of samples should be taken to give the properties of the fluids in different parts of the reservoir. It may not be necessary to run complete PVT analyses on all samples; but key determinations, together with producing gas-oil ratio surveys, temperature surveys, tank oil gravity surveys, and, in the case of an undersaturated fluid reservoir, a saturation pressure survey, can be used to define the "average" fluid properties.

(2) Samples should be taken during the early life of the reservoir. It is not too difficult to estimate the properties of the original reservoir fluid from samples taken while the reservoir pressure is still above 80 per cent of its original value. It is almost impossible to do so accurately with samples taken after the reservoir pressure has declined to 40 per cent of its original value.

(3) Samples should be re-run after the reservoir pressure has changed materially. Laboratory tests are made on the basis of specific test processes and the assumption that thermodynamic equilibrium exists between phases at all times. The estimation of the fluid properties in a reservoir at, say, 1000 psi can be made more accurately from samples obtained when the reservoir is near 1000 psi than from samples taken when the reservoir pressure is 3000 psi.

Three types of samples are commonly used in determining properties of reservoir fluids: first, bottom-hole samples from near the bottom of the tubing, taken either while the well is flowing at a slow rate or shut in; second, recombined samples of the liquid and gas from a surface trap; and third, surface flow-line samples ordinarily obtained at the well tubing head. The remainder of this chapter will touch on several points in favor of, or limiting, the accuracy of these three types of samples and will outline methods by which the samples are taken. The discussion of bottom-hole sampling methods pertains to dissolved gas systems; recombination samples will be discussed from the viewpoint of both dissolved gas and condensate systems. Flowline samples will be considered entirely in connection with condensate systems.

Bottom-hole Sampling Equipment

Most samplers are designed to be run inside 2-inch tubing on a steel wire of the type used in the operation of pressure recorders. The samplers are of the order of 6 feet in length and can hold approximately 600 cc of fluid.

In order to get a sample of the fluids near the bottom of the well, a trip valve mechanism is provided for opening and closing the fluid ports of the sampler. Samplers differ mainly in the type of valve mechanism used for trapping the sample. The Humble Type A sampler operates when a sudden movement of the sampler causes a suspended weight to drive a hardened steel point through a thin steel disk, allowing the fluid in the well to enter the previously evacuated sample chamber. The Humble Type B sampler also allows well fluids to enter a previously evacuated chamber, but the operation is carried out by the temperature and pressure at the sampling depth, which cause rupture of a tin disk. In both types of Humble samplers a second check valve seals the fluid entry port when the sampler is withdrawn from the well.

The Bureau of Mines, the Perco, and the Wofford type samplers differ from the Humble samplers in that a port in each end of the sampler remains open while the sampler is lowered into the well.* Thus, fluid can flow through the sampler while it is being run into the well and until the fluid ports are caused to close. The valves closing the fluid ports can be set in operation by either a clock-electromagnetic time latch or a shear pin arrangement that operates when a sudden movement is imparted to the sampler. Clock-operated mechanisms seem to be the most successful, the chief disadvantage being that the clock almost doubles the cost of the unit.

From the above brief descriptions, it can be appreciated that a bottom-hole sampler might fail to operate for one or more of the following reasons:

* More detailed description of the Humble type sampler has been published by Schilthuis.[5] Grandstone and Cook[6] have described the Bureau of Mines sampler. Catalogs of the Engineering Laboratories, Inc. describe the Perco and Wofford type samplers.

(1) The thermo-pressure disk type may not rupture, or may rupture prematurely if the disk is not of the proper thickness for the pressure and temperature conditions in the well.

(2) Mechanisms operating on the principle of a sudden movement of the sampler may function prematurely when running the sampler to depth. This often is caused by striking a liquid surface or running into a paraffin accumulation in the tubing.

(3) Sudden movement mechanisms may fail to operate when used in wells containing a high viscosity oil or in deep wells. This is because sufficient movement cannot be imparted to the sampler by jerking the supporting line at the surface.

In addition to the bottom-hole sampler, considerable surface equipment is necessary to make spot field tests on the sample obtained in the well and for transferring the sample from the bottom hole unit to containers for transportation to the laboratory.

Plan of the Bottom-hole Sample Method

Several reasons were listed early in this chapter that cast doubt on our ability to reconstruct in the laboratory an exact duplicate of a reservoir fluid. Of course, a bottom-hole sample obtained in the tubing under a pressure such that only a single fluid phase could exist would be representative of the fluid in the tubing and probably representative of the fluid in the vicinity of the well.

When a well is flowing and both gas and liquid phases exist at the point of sampling, it is difficult to say whether the gas and liquid flowing past the sampler is mixed homogeneously. If the gas is dispersed homogeneously throughout the liquid, then the sampler should obtain a representative sample of the material flowing in the tubing. However, the well might "head," which can result in periods during which the material flowing past the sampler may be either predominantly gas or liquid. Certainly, a sampler of the Humble type is filled within a few seconds after the disk is ruptured and samples whatever happens to be flowing past it during that short interval. While it is more difficult to picture the actions within, say, the Perco sampler, it is reasonable to assume that the proportion of gas and liquid in the sampler is unknown at the instant the valve mechanism operates to close the sampler. Consequently, many bottom-hole samples should not be considered as representative of any specific fluid in the well or the reservoir. Rather, they are better considered to be only "samples" that have a limited chance of being a replica of the reservoir fluid.

Fortunately, one thing can be done toward limiting the range of variation of fluid composition obtained in the sampler. By sampling the well shortly after it has been shut in, there exists a smaller chance of forming the sample from indeterminate amounts of gas and liquid. If low pressure draw-down conditions are established in the well prior to sampling, and if after shutting in the well sufficient time elapses for any free gas to segregate in the top of the tubing, the sampler usually obtains a completely liquid sample, saturated at previously flowing bottom-hole pressure and temperature. Such a sample is reproducible, and a second or third sample will exhibit bubble point pressures and other properties in good agreement with those of the first sample. As will be shown in Chapter 7, it is possible to estimate the properties of the reservoir liquid phase by extrapolating the measured properties of the bottom-hole liquid sample.

Bottom-hole Sampling Methods

In this section a short description will be given of the field operations of taking bottom-hole samples. The purpose is not to present a detailed description of the various steps necessary for the preparation of the sampler and the sampling itself, but rather to discuss the reasons for the various tests made in the field.

An obvious precaution, but one frequently overlooked, is to determine whether free water exists in the well at the desired sampling depth. The presence of both water and hydrocarbon liquid in the sampler cannot be determined easily without taking a chance of destroying the sample. As it takes considerable time to run the sampler to depth, take the sample, bring it to the surface, and determine the probable contents, the simple precaution of running a tubing bailer to depth and looking at the liquid brought to the surface is well worth the extra effort.

Between the time that the sampler traps a sample of fluid and the time the sample is studied in the laboratory, several simple tests can be made in the field to identify partially the material obtained in the sampler and to determine whether fluid loss occurs while the sample is in transit to the laboratory. These tests are:

(1) To determine the pressure existing within the sampler when it is received at the surface.

(2) To determine the compressibility of the material within the sampler.

(3) To determine the bubble point pressure of the material in the sampler.

In Figure 8 is shown the arrangement of a calibrated mercury injection pump, pressure gauge, and

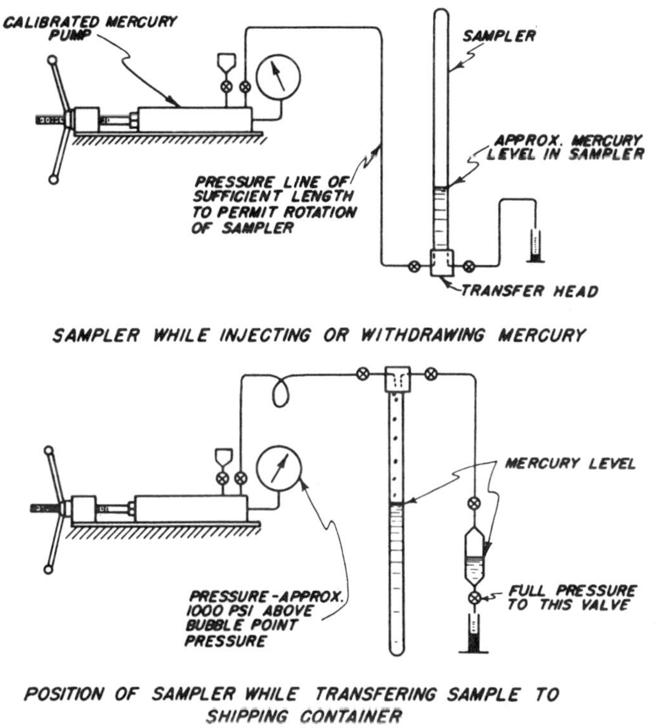

Figure 8—Arrangement of Sampler and Auxiliary Equipment While Determining Probable Contents and Transferring Sample

sampler used in field testing the sample and in transferring the sample to shipping containers. As the sampler is brought to the surface, the pressure exerted by the fluid inside keeps the valves in the instrument closed. After the sampler is removed from the well, a transfer head assembly is connected to one end and mercury from the mercury injection pump is forced against the sampler valve. When the mercury pressure is raised to the point where it is greater than the pressure of the material within the sampler, the valve is forced open. Mercury or sample material can then be injected into or withdrawn from the sampler, as shown in Figure 8. The pressure existing inside the sampler when the valve first opens is roughly indicative of the material within the sampler. For example, if the sampler were completely filled with liquid saturated at bottom-hole conditions of, say, 2500 psi and 200°F, the pressure existing in the sampler at 80°F would be near 1650 psi. The reason for this is that contraction of the liquid when cooled to 80°F causes a reduction in pressure, which is partially offset by gas coming out of solution from the liquid. Had a sample of gas been obtained at 2500 psi and 200°F, the pressure in the sampler at 80°F again would be near 1650 psi. However, had the sampler obtained a sample of water at the bottom-hole conditions, the low solubility of gas in water would cause the sample pressure at 80°F to be much lower, probably close to atmospheric. Thus, the pressure in the sampler measured when the sampler valve first opens gives some indication of a water sample. By itself, it usually cannot be used to differentiate between a gas sample and a saturated liquid sample, or combinations of gas, saturated liquid, and water samples.

The compressibility of the material in the sampler also is roughly indicative of the material sampled. For example, if the sampler obtained 600 cc of saturated liquid at 2500 psi and 200°F, the 600 cc of liquid would form approximately 50 cc of gas and 550 cc of liquid when cooled to 80°F. If the gas were forced back into solution at 80°F by injecting mercury into the sampler, the pressure in the sampler would increase at about 40 psi per cc of mercury injected. After the bubble point pressure at 80°F was reached, further injection of mercury would cause an apparent compressibility of about 300 psi per cc of mercury injected. If the sampler had obtained a sample of gas, the greater compressibility of gas would result in a rate of pressure increase of about 3 psi per cc of mercury injected.

If water constituted the fluid obtained by the sampler, the pressure would not increase appreciably above atmospheric until about 15 cc of mercury had been injected. The pressure then would increase at a rate of about 1000 psi per cc of injected mercury.

The measured bubble point pressure, when corrected to the bottom-hole temperature,* also is a guide to the probable sampler contents. A corrected bubble point pressure lower than the bottom-hole pressure may indicate the reservoir liquid to be undersaturated, or it may indicate leakage of gas from the sampler as it was raised to the surface. A bubble point pressure higher than the bottom-hole pressure indicates that both gas and liquid existed in the sampler at the time the sampler valves closed, or that liquid phase has leaked from the sampler.

Unfortunately, it is impossible to list numbers or values for the three tests which will identify accurately the material obtained in the sampler or indicate whether the sample is a desirable one for laboratory work. Variations in the properties of crude oil and in the quantity of gas in solution under the wide ranges of pressure and temperature encountered in oil fields cause differences in the values of the numbers indicated above. Consequently, the best criterion as to whether a "good" sample has been obtained is if a second or third sample taken from the well immediately afterward has the same measured properties as the first. An illustration of this criterion is shown in Figure 9, in which duplicate samples taken in two adjacent wells producing undersaturated liquid samples gave comparable bubble point pressures. Roughly the same amount of mercury was injected to reach the bubble point pressure, and the initial pressures in the sampler were equivalent. These data would indicate that any of the four samples could be used for laboratory tests.

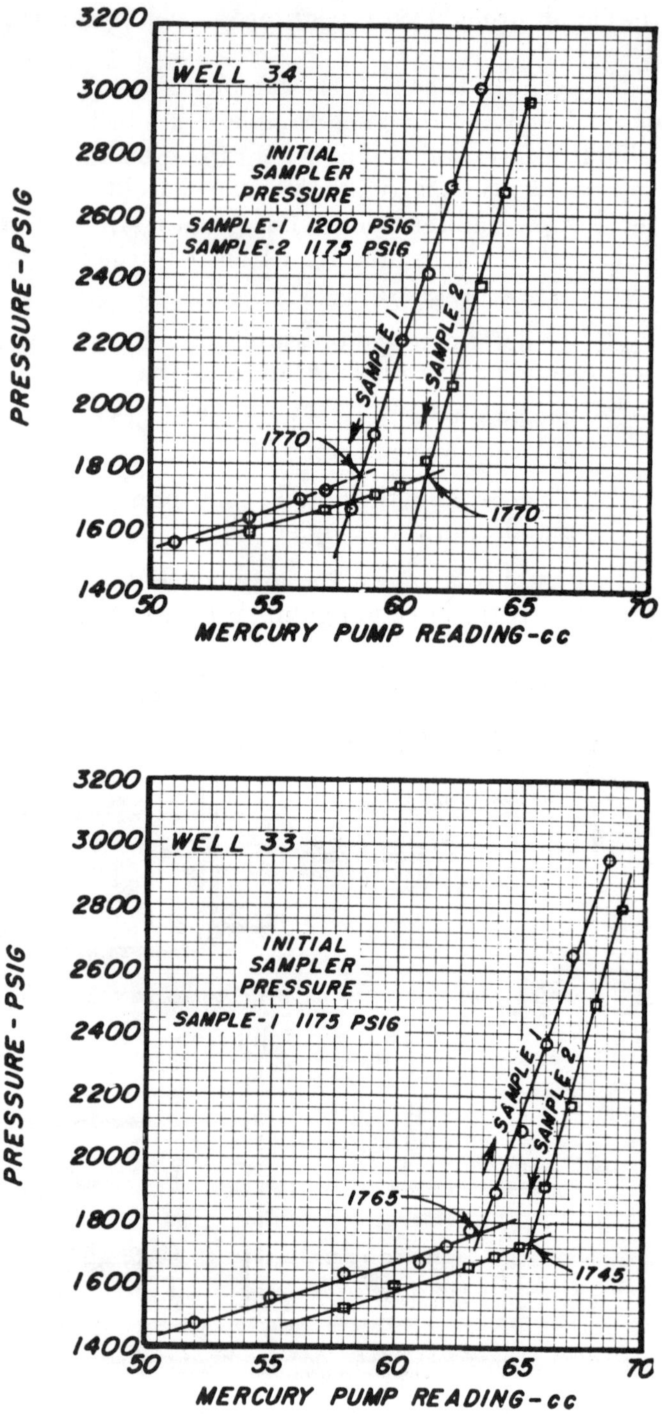

Figure 9—Results of Field Test on Four Bottom-hole Samples

Plan of the Recombined Surface Samples Method

By recombining samples of all surface gases and liquids in the proportions in which they are produced from a well, a hydrocarbon system is obtained that is representative of the fluid produced at the well head. Under steady state flow conditions in the well tubing, the fluid compositions are the same at the well head and at the bottom of the tubing. Thus, a true reproduction of the fluid entering the well tubing can be obtained by proper recombination of surface samples. This, of course,

* Bubble points increase at about 500 psi per 100°F increase in temperature. This is not true of all systems, however.

does not necessarily give a true reproduction of the reservoir fluid as a whole, but rather only a hydrocarbon system that can be used in estimating the reservoir fluid properties.

As a matter of laboratory operational convenience, samples used in recombination methods are usually from the primary trap. The recombination of these samples to reproduce the well effluent hinges on accurately metering the two streams produced from the primary trap. Most field installations do not provide for metering trap liquid directly, and, as a consequence, the trap liquid flow rate must be calculated from the tank oil production rate and from a shrinkage factor to compensate for the decrease in volume of liquid as it progresses from the trap to the tank. While metering of primary trap gas and the gauging of tank oil are field operations, the determination of the approximate shrinkage factor is usually a laboratory measurement.

Accurate gas and oil production rates are important in picturing the state of the fluids in a reservoir on the basis of recombined sample tests. For example, a bubble point pressure-gas-oil ratio relation can be determined in the laboratory with probable errors of a few pounds per square inch and one or two cubic feet per barrel. The use of such a relation, however, to decide whether the well effluent exists as one or two phases under reservoir pressure and temperature conditions depends on field gas-oil ratio measurement. Unfortunately, field measurements are often in error by more than 5 per cent. An error of this magnitude can easily cause a 150 psi error in defining the bubble point pressure of the well effluent. Thus, the recombination method is inferior to the bottom-hole method, primarily when an accurate measure is desired of the bubble point pressure of an undersaturated reservoir fluid.

Before passing to a discussion of surface sampling methods, a word should be said regarding the recombination of surface trap samples in ratios other than their produced ratio. The purpose of such tests are two-fold: First, data can be obtained on the physical properties of a wide variety of systems. Bubble point pressures, dew point pressures, formation volumes, and partial phase diagrams can be correlated in terms of gas-oil ratio, gas gravity, oil gravity, pressure, and temperature to give generalized PVT relations for use where specific data are not available.[7,8] Secondly, data obtained on other than produced ratio systems often can be used in calculations pertaining to past or future behavior of fluid from a particular well or field. For example, for economic reasons laboratory tests are often deferred until the reservoir pressure has declined from its original value and an indication of the reservoir size has been obtained. Over limited reservoir pressure and producing gas-oil ratio ranges, gas and liquid sampled at the primary trap will not differ greatly from the gas and liquid originally produced from the trap. Consequently, a reasonable reconstruction of the original produced system can be made by recombining the later date samples in the original produced ratio.

Surface Sampling Methods

Field trap installations seldom permit a choice of sampling points from which to obtain gas and liquid samples. Ordinarily a valve for sampling gas is provided at the location of the gas meter. Liquid samples as a rule must be taken from the bottom of the liquid level glass, although this location is less desirable than at an outlet directly ahead of the oil dump valve.

A simple gas sampling scheme is shown in the upper part of Figure 10. A ¼-inch copper tube connects a gas sampling point to the bottom of a drying tube containing 4 to 6-mesh calcium chloride. The sample container for collecting and storing the sample is connected to the top of the drying tube. The drying tube is maintained under full trap pressure, and the quantity of gas admitted to the sample container is controlled by means of the valve at the top of the dryer. After purging the sample container several times to remove air, the container is filled to the desired pressure for shipment to the laboratory. This method works well for relatively dry gases and usually can be relied upon to give a representative gas sample where the trap pressure is less than 500 psi and the trap temperature is near or below atmospheric temperature. At higher trap pressures, or where the trap temperature is considerably higher than atmospheric temperature, the gas may contain a large proportion of heavy compounds. The sudden reduction in pressure when the gas first enters the container may cause a retrograde liquid to form. The accumulation of any retrograde liquid in the container during the purging process will give a nonrepresentative sample. Where this condition might exist, one should dispense with the drying tube and take the sample by displacing water from the container under the full trap pressure. However, this second method is not too satisfactory if the trap gas contains a quantity of water-soluble constituents, such as H_2S and CO_2.

The sampling of trap liquid is illustrated in the lower part of Figure 10. Liquid flows upward

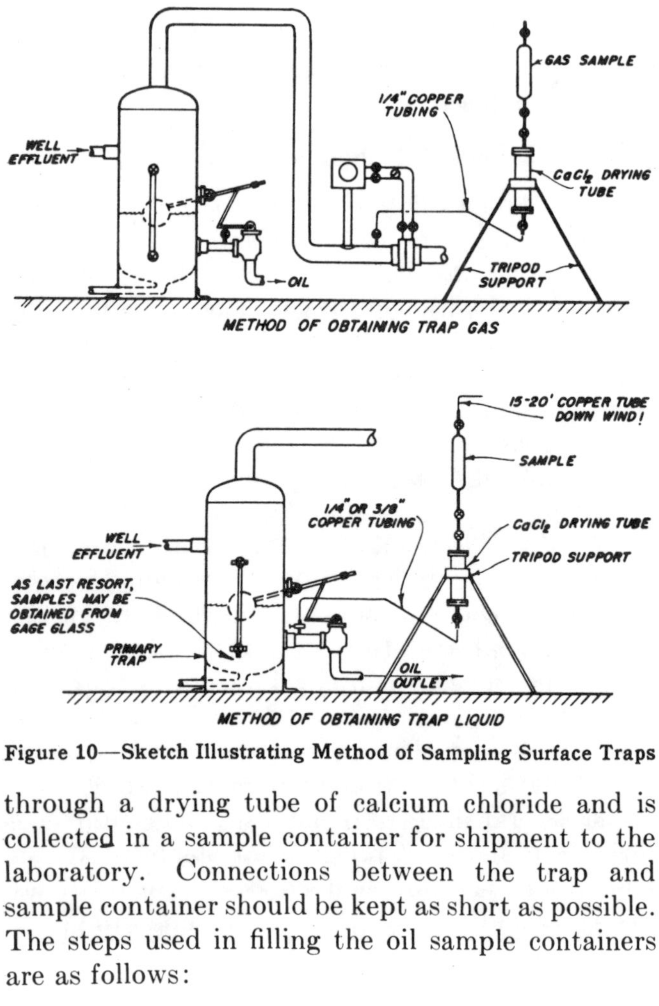

Figure 10—Sketch Illustrating Method of Sampling Surface Traps

through a drying tube of calcium chloride and is collected in a sample container for shipment to the laboratory. Connections between the trap and sample container should be kept as short as possible. The steps used in filling the oil sample containers are as follows:

(1) The sample container is connected to the top of the drying tube, leaving the top valve of the container closed and the bottom valve open.

(2) The valve at the top of the drier is opened slowly to admit liquid into the sample container. The container is filled to full trap pressure. This step must be done slowly so that gas is not caused to come out of solution in the oil in the drier.

(3) The air originally in the sample container, any gas that came out of solution as the liquid first entered the container, and the flashed liquid are bled out through the top valve of the container. At least three container volumes of new liquid are then bled through the sample container in order to displace all liquid that was altered by the original flashing process.

(4) Both valves of the sample containers are closed, and the container is removed and shipped full of liquid.

A second method of obtaining a liquid sample is to displace water from the sample container under full trap pressure. The drying tube is dispensed with in this method. Only about 90 per cent of the water should be displaced by the oil sample if the temperature of the sample is likely to increase before reaching the laboratory. The final 10 per cent of water should be removed after the valve between the trap and the container is closed. This creates a small gas cap within the container which will prevent large increases in pressure for unavoidable increase in temperature.

The design of satisfactory sample containers is shown in Figure 11. It has been found that 500-cc

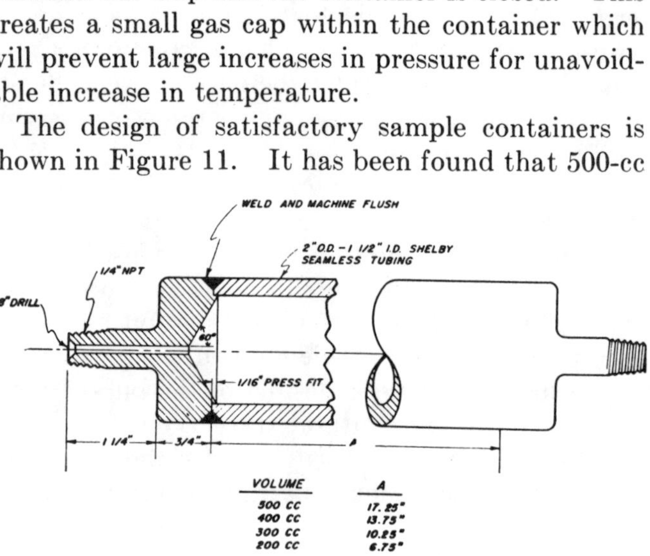

Figure 11—Diagram of Liquid and Gas Sample Containers

gas containers and 50- to 100-cc liquid containers are the most useful sizes for laboratory work.

The best rate at which to flow the well prior to and during the sampling period has to be considered individually for each well. In general, the best producing rate is the one which results in the smoothest well operation and the best measurement of surface products, but which is still consistent with the normal field operation of the well. The factors usually considered when selecting a producing rate are primarily of a mechanical nature; they include whether the well flows by heads when producing at certain rates, the size of surface-gas metering facilities, and the size of oil gauge tanks. These factors influence the rate selection in the following manner.

Flow Characteristics of the Well. The alternate slug flow of liquid and gas into the surface traps may cause minor changes in the composition of the trap samples, but a more important consideration is the difficulty in determining an accurate gas rate during the observation period. For this reason, the well should always be produced at a rate that will result in a smooth flow of gas and liquid from the surface trap.

Surface Meter Installations. When an effort is made to flow the well at low rates, the quantity of primary trap gas may be so small that the gas cannot be metered accurately. One important factor in the selection of the producing rate is the size of meter runs installed at the well and whether the available orifice plates will give good, reliable meter charts. In some instances, it may be necessary to lower the trap pressure in order to increase the accuracy of the gas metering. However, this procedure is recommended only as a last resort, because changing the primary trap pressure will, in itself, result in a different produced gas-oil-ratio. Even though the PVT tests may be extensive enough to evaluate the effect on producing gas-oil ratio of changing the trap pressure, the changing of trapping conditions from the normal field conditions unnecessarily complicates the use of the laboratory data.

Stock Tank Installations. Field stock tank installations are not always the best size for determining short time production rates because they are built primarily for accumulating oil for shipment. However, plotting the cumulative oil production vs. time usually will give a satisfactory production rate figure.

It is advisable to flow a well for an extended period before taking samples. Three to four days usually is adequate, provided the flow rate and producing gas-oil ratio remain relatively constant during that time. Tubing pressure, casing pressure, and gas and oil production rates should be gauged at 8-hour intervals prior to the sampling period, and the readings should be made at hourly intervals during the sampling period. The flowing bottom-hole pressure should be measured immediately after obtaining the samples.

Flowline Samples

Samples of the gas and liquid flowing in the tubing at the well head or in the flowlines connecting the well with the surface traps are sometimes used for PVT tests. This type of sample is used primarily where a number of condensate wells are producing into a common field collection line or into a single trap installation.[9,10] To sample the production from an individual well, it is necessary to sample upstream of the gathering system. By experience it has been found easier to use a small laboratory-scale trap and sample only a small portion of the well effluent than to move a full-scale trap to the well under test and connect it into the flowline.

The basic sampling device, called a tip, consists of a straight section of small tubing inserted into the well tubing or flowline with its open end pointed upstream to the flow of fluid. The sample entering the tip flows through the tubing to a small-scale trap. An indication of the composition of the sample is obtained by measuring the amount of liquid formed when the sample is brought to a predetermined pressure and temperature in the trap.

Many complex problems are involved in obtaining a representative sample of two-phase well production from the well tubing or surface lines. Any liquid present in the stream wets the sides of the flowline and forms a film that builds up on the pipe wall and flows along with the gas phase. The gas flowing past this film tends to ripple its surface and tear off droplets; consequently, the liquid phase moves both as film on the pipe wall and entrained liquid in the gas stream. The distribution of the liquid is an unknown function of the turbulence in the stream.

When the sampling tip is placed in the center of a flow stream, it cannot by itself sample aliquot parts of the film liquid, entrained liquid, and gas. However, the velocity of the sample stream entering the tip can be controlled by regulating the back pressure at the downstream end of the sampling tube, and can be made either greater or less than the fluid velocity in the body of the flow stream. As illustrated in the upper portion of Figure 12, the sample tip can be caused to take various proportions of entrained liquid and gas by controlling the sampling rate. This is because the inertia of the more dense, dispersed liquid droplets causes them to maintain their motion and velocity and enter the sampling tip in increasing proportions as the sampling velocity decreases. When the ratio of fluid velocity in the sample tip to the fluid velocity in the flow section is plotted against the correction factor necessary to make the sample composition agree with the flow stream composition, a curve is obtained similar to the one illustrated in the lower part of Figure 12.

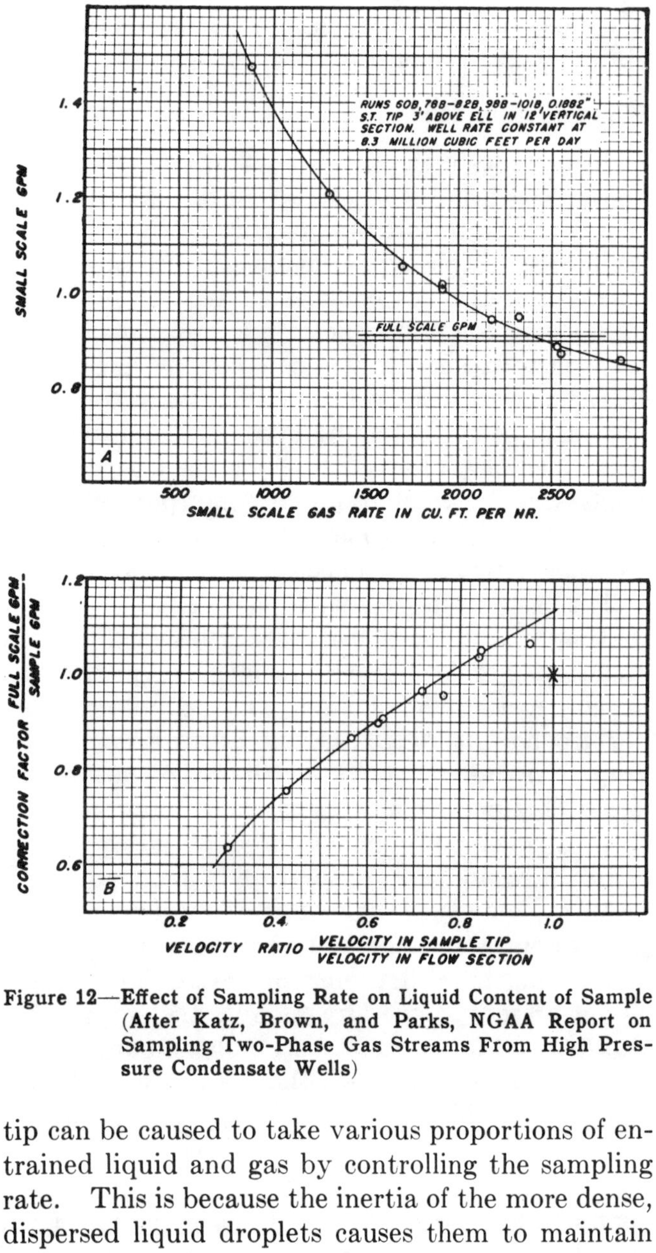

Figure 12—Effect of Sampling Rate on Liquid Content of Sample (After Katz, Brown, and Parks, NGAA Report on Sampling Two-Phase Gas Streams From High Pressure Condensate Wells)

An ideal sampling device would be insensitive to velocity ratio. Preferably, it also would have a unity correction factor. However, it cannot exist for a flow stream in which one of the phases preferentially wets the walls of the pipe. Therefore, the problem of obtaining samples from flowlines is primarily that of calibrating and using a particular tip under controlled conditions of tip location, sample velocity, and flowline velocity.

Obtaining a representative sample from the well is not, of course, the whole problem in determining the PVT behavior of the well effluent. In order to operate with flowline samples, it is almost mandatory that the tests be of the dynamic type, wherein the sample continuously flows through the small-scale trap, controlled as to its pressure and temperature. This usually requires a mobile laboratory so that the tests can be run near the well location.

TABLE 2. SUMMARY OF ADVANTAGES AND DISADVANTAGES OF VARIOUS SAMPLING METHODS

ADVANTAGES	DISADVANTAGES
Bottom Hole Samples	
1. Often useful to gauge the gas in solution in reservoir oil where accurate gauging facilities are not available.	1. Chance of leakage before sample is removed from well.
2. Works well on undersaturated liquids, giving accurate measure of saturation pressure.	2. Does not, except in case of undersaturated oil, obtain a true sample of reservoir liquid.
	3. Sampler is costly and is subject to loss in well.
	4. Use apparently is limited to dissolved gas systems. Is not used to sample condensate systems.
Recombined Surface Trap Samples	
1. Less expensive equipment.	1. Proper proportions to recombine are dependent on field gauging. May not, even in undersaturated fluids, give accurate measure of bubble point pressure or dew point pressure.
2. Easy to study mixtures of differing compositions (gas-oil ratio).	
3. Samples are easy to handle in laboratory.	
4. Can be used in condensate field studies.	
Surface Flowline Samples	
1. Can be used in condensate fields where separate trap settings are not used.	1. Tip requires calibration for the specific conditions of test.

Summary of Principal Points

Each of the three sampling methods reviewed in this chapter has its strong points. No one sampling method is applicable under all conditions, although the recombined surface sample method has the widest applicability. A summary of advantages and disadvantages of the three methods is presented in Table 2.

Because of the complex fluid behavior encountered in reservoir rocks and because of the many ways that reservoir fluid can differ in composition within different areas within a reservoir, one is seldom justified in directly applying the results of a single PVT determination to the reservoir fluid as a whole. Rather, it should be recognized that the limitations imposed by sampling in the well bore or at the surface usually require extrapolation of the PVT data to reservoir fluid conditions.

Two or three trips with a bottom-hole sampler should always be made before any one of the samples obtained at the surface can be assumed to be representative of the fluid at the well bottom. The fluid remaining in the sampler after part of the original fluid has leaked out while the sampler was being withdrawn from the well is of little use in estimating the properties of the reservoir fluid. While the final laboratory PVT report seldom includes the results of the tests made in the field to identify the sampler contents, it is in order for the engineer to request these data to assure himself that the fluid whose properties have been measured in the laboratory is the same fluid sampled at the well bottom.

The recombined surface sample method at times suffers inaccuracies because of its dependence on field measured gas-oil ratios to reconstruct the well effluent. In many instances the uncertainty in the use of the PVT data caused by the measured field gas-oil ratio applies to an equal degree to bottom-hole samples. For example, the evaluation of the fraction of produced surface gas that exists as free gas in the reservoir rock involves the difference between the field measured producing gas-oil ratio and the laboratory measured dissolved gas-oil ratio. The laboratory measured solubility can be equally accurate whether determined from bottom-hole or recombined surface samples.

CHAPTER 3

BEHAVIOR OF GASES

Almost all petroleum engineering problems use the PVT behavior of gases in one form or another. Gas reserves, gas pressure gradients in well bores, and the calculation of gas flow rates at the surface are three example problems requiring a knowledge of gas behavior.

The basic equation expressing the PVT behavior of gases is the ideal gas law,

$$P\bar{V} = \frac{RT}{M} \quad (3)$$

where, in usual engineering units,
P = pressure, absolute (psia)
$\bar{V}$ = specific volume of the gas (cu ft per lb)
T = temperature, absolute, °R = (460 + °F)
R, arbitrary constant, = 10.73 for all gases
M = molecular weight of the gas

Thus, as expressed by Equation (3), the volume of a unit weight of an ideal gas at any equilibrium temperature and pressure can be evaluated solely from a knowledge of the molecular weight of the gas.

If Equation (3) is changed to represent the PVT state of m pounds of gas by multiplying both sides by m, we have

$$PV = \frac{mRT}{M} \quad (4)$$

where V = total volume occupied by m pounds of gas (cu ft)

The number of unit pound molecular weights, N, comprising the system is represented by the term m/M in Equation (4). In practice, N, referred to as the pound moles or moles of gas comprising the system, can be evaluated from the relation $N = m/M$, or it can be evaluated in terms of a volume of ideal gas measured at a reference pressure and temperature. An example of the foregoing is that 28.97 pounds of air represents one pound mole of air, just as 16.04 pounds of methane represents one pound mole of methane. If either 28.97 pounds of air or 16.04 pounds of methane is confined at 14.7 psia and 60°F (called standard pressure and temperature conditions), it is found that it occupies very close to 379 cu ft of space. In fact, if they were "ideal" gases, they would occupy 379.4 cu ft per lb mole at 14.7 psia and 60°F.

In practice, it has been found that no gas obeys this simple law over all pressures and temperatures. Hydrocarbons, in particular, deviate widely from the ideal gas law at the elevated pressures and temperatures of petroleum reservoirs. In order to express a more exact relationship between the variables, P, V, and T, a correction factor must be introduced into Equation (4). Either of two methods is normally used; the first method introduces a multiplying factor into the right side of Equation (4), and the second method makes use of an adding factor. The multiplying factor goes by several names; three of the most common are compressibility factor, deviation factor, and super-expansibility factor. The compressibility factor and deviation factor are synonymous, and the super-expansibility factor is the reciprocal of these two factors. The adding factor is termed the residual volume. Both methods are accurate to the same degree, but the first is easier to use and for this reason is the most popular.

The Compressibility Factor

Placing the compressibility factor, Z, in the right side of Equation (4) and at the same time substituting N for m/M gives

$$PV = NZRT \quad (5)$$

Equation (5) is the basic equation for gas behavior calculations. An illustration of its use is given in

Table 3. The compressibility factor of an ideal gas is unity at all pressures and temperatures. The compressibility factor of a dry natural gas at atmospheric pressure and temperature is about 0.998 and is ordinarily assumed to be unity. At 2000 psia and 60°F, however, the compressibility factor of a dry gas may be as low as 0.63 and increase to as high at 1.58 at 10,000 psia and 60°F.

TABLE 3. CALCULATION OF MASS AND VOLUMETRIC FLOW OF METHANE IN A PIPELINE

Problem: To calculate the mass flow rate and the standard cubic feet per day of methane gas flowing in a 6-inch I.D. pipeline operating at 1000 psia and 140°F, assuming the average linear flow velocity in the pipe to be two feet per second.

Solution: Area of pipe $= \frac{\pi}{4}\left(\frac{6}{12}\right)^2 = 0.1965$ sq ft

Volumetric rate = 0.1965 · 2 = 0.3930 cu ft per second

From Equation (5),

$$PV = NZRT = \frac{mZRT}{M}$$

At 1000 psia and 140°F, $Z = 0.94$ (Figure 13)

Molecular weight, M, = 16.04

$$m = \frac{PVM}{ZRT} = \frac{1000 \cdot 0.3930 \cdot 16.04}{0.94 \cdot 10.73 \cdot (460 + 140)}$$

= 1.043 lb per second

Mass rate = 1.043 · 3600 · 24 = 9.01(10⁴) lb per day

16.04 lb methane = 1 lb mole = 379 cu ft at 14.7 psia and 60°F

$$\text{Standard cubic feet per day} = \frac{9.01(10^4) \cdot 379}{16.04}$$

= 2.13(10⁶) cu ft per day

The compressibility factors of different gaseous compounds vary over wide ranges at elevated pressures. For example, at 140°F, the compressibility factors of methane, ethane, and propane vary with pressure as shown in Figure 13. It can be seen that a considerable difference exists in the compressibility factor behavior of these first three members of the paraffin hydrocarbon series. Wide differences would also be expected for the higher members of the series.

During the 1860's, many papers appeared in the scientific literature concerning the PVT behavior of individual gases. J. D. van der Waals[11] in 1873 developed the theorem of corresponding states on which the modern correlation of compressibility factors is based. The theorem proposes that all gases will exhibit the same behavior when viewed in terms of reduced pressure, reduced volume, and reduced temperature. The term "reduced" means that each variable is expressed as a ratio of the critical value. Thus, all variables are reduced to a common datum, the value of the critical. The reduced conditions of volume, temperature, and pressure are expressed as

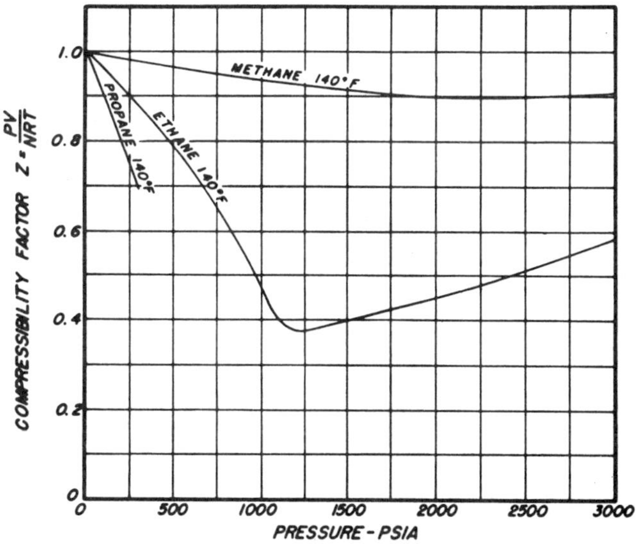

Figure 13—Compressibility Factors of Methane, Ethane, and Propane at 140°F as a Function of Pressure

$$\bar{V}_R = \frac{\bar{V}}{\bar{V}_C}, \qquad T_R = \frac{T}{T_C}, \qquad P_R = \frac{P}{P_C} \tag{6}$$

where subscript $_R$ indicates the reduced state, and subscript $_C$ indicates the critical state.

The theorem of corresponding states is not perfect. However, when applied to gases having similar chemical structure (as paraffin hydrocarbons), it offers a correlation method with satisfactory accuracy for engineering work. This is shown by the reduced PVT data on methane, ethane, and propane plotted in Figure 14. The curve in Figure 14 is an average curve sketched in to show the degree of correlation.

In 1936, W. B. Kay[12] presented experimental data and outlined a correlation method that can be used for mixtures of gaseous compounds. Kay's work forms the basis for the modern methods concerned with natural gases. It can be summarized as follows:

(1) The theorem of corresponding states, with modifications, can be applied to a mixture of gases, as well as to individual gases.

(2) In order to correlate the PVT behavior of a gas mixture with the reduced states of individual gases, it is necessary to use fictitious values for the critical temperature and pressure of the mixture. The fictitious values are

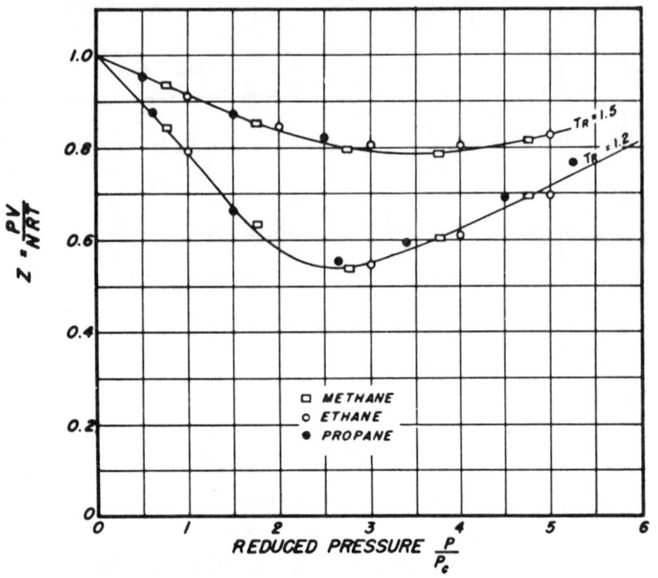

Figure 14—Compressibility Factors of Methane, Ethane, and Propane as a Function of Reduced Pressure and Reduced Temperature

called "pseudo" critical values to differentiate them from the true critical values of the mixtures.

(3) The pseudo-critical pressure and temperature of a mixture of low molecular weight gases are obtained by summing the product of the mole fraction of each component times the component's true critical value. For example, the pseudo-critical temperature of a mixture composed of 80 mole per cent methane and 20 mole per cent propane is $0.80 \cdot 344 + 0.20 \cdot 666 = 408.4°R$.

(4) High molecular weight compounds or mixtures of compounds require pseudo-critical rather than true critical values when used in the theorem of corresponding states. The correct pseudo-critical constants are evaluated by determining the values which give the best agreement between experimentally determined volumetric data and calculated volumetric data. For example, Kay found that for a binary mixture of 26.62 mole per cent ethane and 73.38 mole per cent *n*-heptane, it would be necessary to use pseudo-critical constants of 990°R and 355 psia for *n*-heptane in order for the theorem of corresponding states to apply to that mixture. The true critical values of *n*-heptanes are 973°R and 397 psia.

Most natural gases handled in surface equipment contain only small fractions of hexane and heavier compounds. Because of the small quantities of heavy components, the calculated pseudo-critical constants of the mixture are only slightly in error when the true critical constants of the heavy components are used in place of proper pseudo-critical values. Furthermore, gas analyses seldom determine properties from which the pseudo-critical values can be determined for the "plus" component. Consequently, the usual practice is to use the true critical constants for all compounds except the plus component and to use the true critical value of the next highest compound to represent the plus component. This is illustrated in Table 4 wherein the critical constants of *n*-octane are assigned to the heptanes plus fraction.

The example in Table 4 illustrates also the method of evaluating a compressibility factor of a gas at an elevated pressure and temperature when the composition of the gas is known. Multiplying the mole fractions of the individual compounds by their true critical values gives the pseudo-critical constants of the mixture. By using the reduced relationship shown in Equation (6), it is possible to evaluate the compressibility factor of the gas at any desired pressure and temperature from suitable compressibility factor charts.

The compressibility factor chart[13] shown in Figure 15 is applicable to most gases encountered in petroleum reservoirs. The calculated volumetric behavior of gases containing only minor amounts of nonhydrocarbons can be expected to be accurate within 3 per cent when using calculated pseudo-critical constants and the reduced relations shown in Figure 15.

Pseudo-critical Constants-Gas Gravity Relations

Unfortunately, the composition of a gas from which to compute the pseudo-reduced state is not always available. As gas gravity measured by a Schilling Bottle or Edwards Balance is a common field measurement, it is advantageous to have a method of predicting the compressibility factor solely from gas gravity.

In Table 1, it can be observed that critical pressures decrease and critical temperatures increase as the molecular weights of the hydrocarbons increase.

TABLE 4. CALCULATION OF GAS PRESSURE GRADIENT IN THE ANNULUS OF A WELL

Problem: To compute the pressure gradient in a gas column at 200°F and 3000 psia, given the composition of the gas.

1	2	3	4	5	6	7	8
Component	Mole Fraction n	Molecular Weight of Component M	n M	Critical Temperature of Component T_c	nT_c	Critical Pressure of Component P_c	nP_c
C_1	0.840	16.0	13.44	344	289.0	673	565.3
C_2	0.083	30.1	2.50	550	45.7	709	58.8
C_3	0.042	44.1	1.85	666	28.0	618	25.9
i-C_4	0.007	58.1	0.41	733	5.1	530	3.7
n-C_4	0.014	58.1	0.81	766	10.7	551	7.7
i-C_5	0.005	72.1	0.36	830	4.1	482	2.4
n-C_5	0.004	72.1	0.29	847	3.4	485	1.9
C_6	0.002	86.2	0.17	915	1.8	434	0.9
C_7+	0.003	114*	0.34	1025*	3.1	370*	1.1
	1.000		20.17		390.9		667.7

Pseudo-reduced temperature $= \frac{660}{390.9} = 1.69.$ Pseudo-reduced pressure $= \frac{3000}{667.7} = 4.5$

Compressibility factor = 0.862 (Figure 15) Average molecular weight of gas = 20.17

$$\text{Density} = \frac{1}{V} = \frac{PM}{ZRT} = \frac{3000 \cdot 20.17}{0.862 \cdot 10.73 \cdot 660} = 9.91 \text{ lb per cu ft}$$

$$\text{Gas gradient} = \frac{9.91}{144} = 0.069 \text{ psi per ft}$$

* C_7^+ fraction has been assigned properties of octane.

With the exception of the critical pressure of methane and minor discrepancies caused by the isomeric compounds, there is good correlation between the critical constants and molecular weight of the paraffin series of hydrocarbons found in petroleum reservoirs. As the pseudo-critical pressure, the pseudo-critical temperature, and the average molecular weight of a gas are always obtained by multiplying the appropriate constants of the components by their mole fractions and adding the resultant products, it would be expected that a correlation of the pseudo-critical constants with the average molecular weight of the gas would exist for gases of similar makeup.

A test of this reasoning is shown in Figure 16. The pseudo-critical constants of 71 California natural gases, calculated by the method illustrated in Table 4, have been plotted against their calculated gravity (ratio of the molecular weight of the gas to the molecular weight of air: 28.97). These 71 gases consist of high-pressure trap gases (200–500 psi), low-pressure trap gases (30–100 psi), and tank vapors, although a few condensate well effluents and laboratory mixtures have been included. As can be seen, a good correlation exists between the variables for these particular gases.

Pseudo-critical constants-gas gravity correlations similar to those shown in Figure 16 have been presented by other authors. Mathews, Roland, and Katz[14] have presented data on 25 Oklahoma City (Wilcox) gases that included high-pressure equilibrium vapors, trap gases, plant intakes, and plant residues. The relations obtained for these gases are shown in Figure 17. In addition, these authors found a relation for 150 miscellaneous gases that falls almost exactly on the California gas relation shown in Figure 17. Standing and Katz[13] obtained a slightly different relation; however, their gases differed from the California gases in that the high gravities were caused by quantities of hexane and heptanes plus material in the gas, rather than quantities of ethane and propane. Consequently, their gravity-pseudo-critical relations would not be expected to be the same as those of the California gases.

Before discussing in detail the use of Figure 17, the reason for the favorable correlations should be emphasized. All the gases in the correlations had nitrogen contents of less than 2 per cent. The carbon dioxide content of the gases was, with few exceptions, less than one per cent. Finally, their hydrocarbon compositions followed a definite trend,

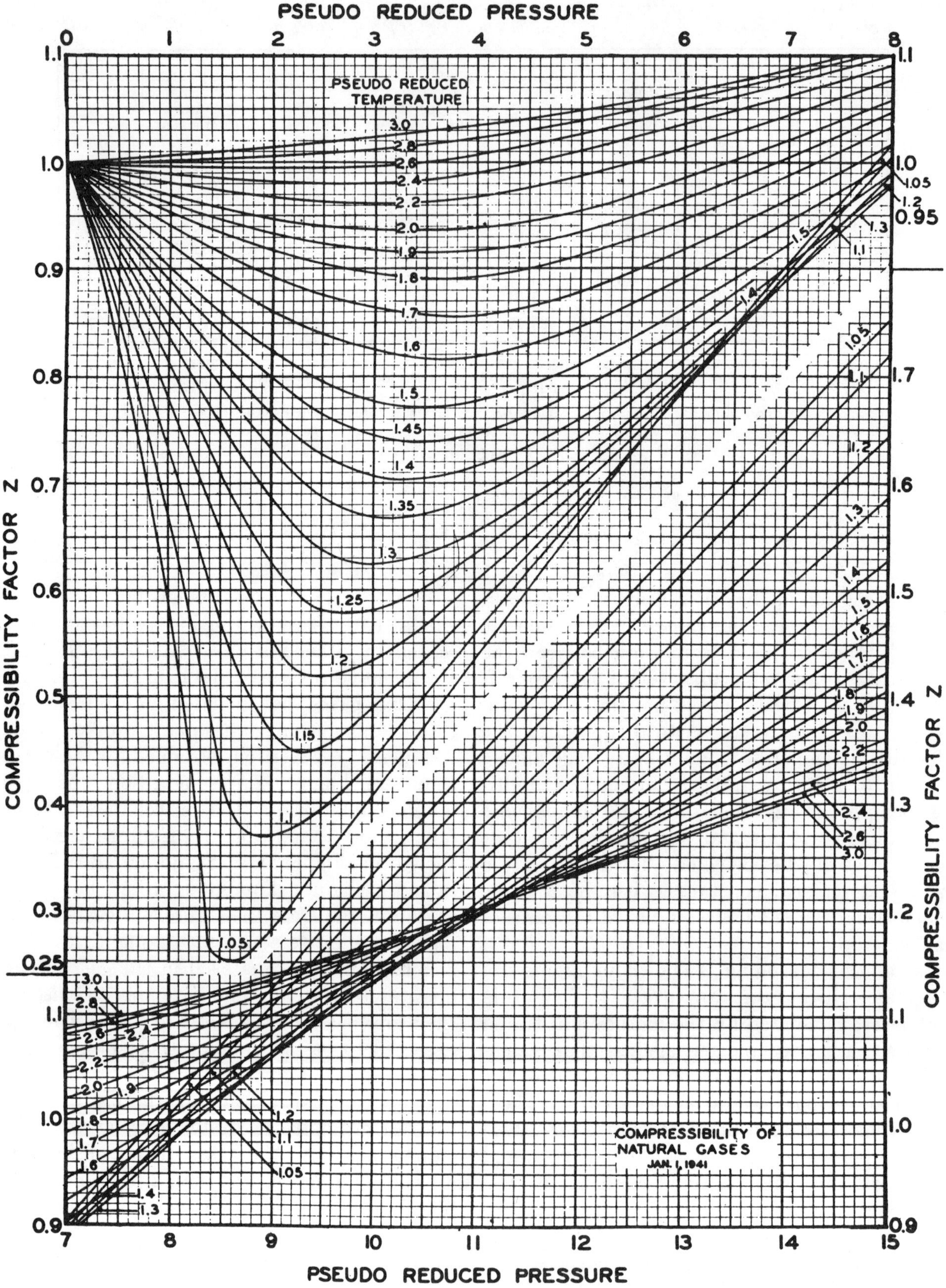

Figure 15—Compressibility Factors for Natural Gases (After Standing and Katz, Trans. AIME, 1942)

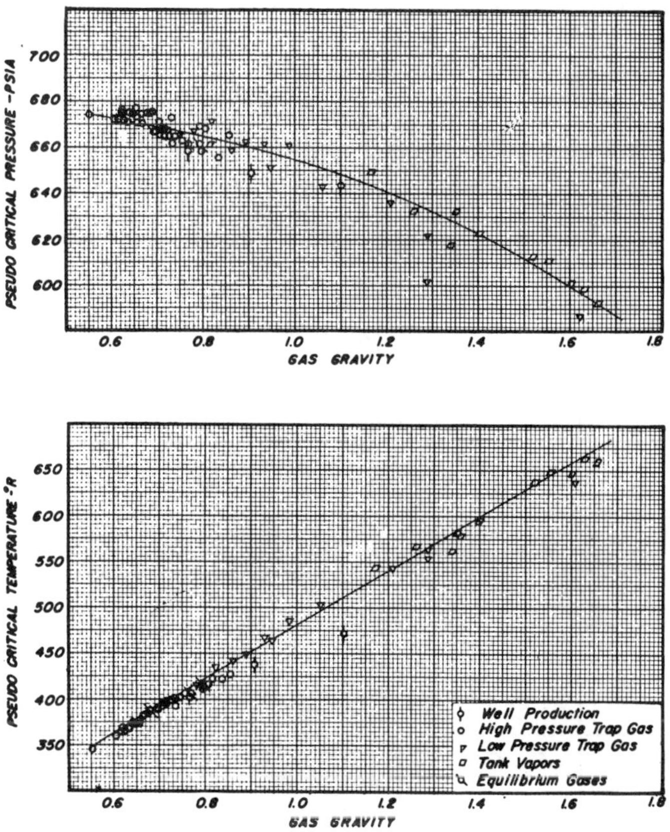

Figure 16—Pseudo-critical Pressure and Pseudo-critical Temperature vs. Gas Gravity. California Natural Gases

i.e., the mole fraction of methane in the high-pressure trap gases and well effluents was of the order of 85 per cent, the ethane about one-tenth of the methane, and the propane about one-half of the ethane. The composition of the gas used in the example in Table 4 is characteristic of the average high-pressure trap gas used in these correlations. The condensate fluids, in addition to having a relative methane-ethane-propane ratio similar to that mentioned, contained up to 5-mole per cent of heptanes plus material.

The relations shown in Figures 15 and 17 greatly facilitate calculations of gas behavior. For example, the pressure-temperature-density relations of gases whose pseudo-critical-gas gravity relations follow the California gases are shown in Figures 18, 19, and 20. These charts, with linear interpolations between charts, give density data which are usually in error by less than 3 per cent. The error to be expected when applying Figures 18 to 20 to an unknown gas can, of course, be greater than 3 per cent, because of the possibility that the unknown gas contains nonhydrocarbon components or a distribution of hydrocarbons different from those used in the correlation. However, as the relations used to obtain Figures 18 to 20 gave good agreement with Mathews, Roland, and Katz's 150 miscellaneous gases, their 25 Oklahoma City gases, and the 71 California gases, some confidence can be placed in the general use of the charts. An illustration of such general use is given in Table 5.

Table 5. Calculation of the Reservoir Volume Occupied by One MCF Surface Gas

Problem: To compute the volume of reservoir space at 2750 psia and 220°F occupied by one MCF of 0.75 gravity surface gas.

Solution: Interpolation at 2750 psia and 220°F between the 0.70 gravity and 0.80 gravity charts (Figures 19 and 20) gives

Density $= 8.77 + 0.5(10.54 - 8.77) = 9.65$ lb per cu ft

$$\text{One MCF surface gas} = \frac{1000}{379} \cdot 0.75 \cdot 28.97 = 57.3 \text{ lb}$$

$$\text{Volume} = \frac{57.3}{9.65} = 5.94 \text{ cu ft}$$

Alternate Solution:

From Figure 17,

$pT_C = 407°\text{R}$ $\quad pP_C = 667$ psia

At 2750 psia and 220°F

$$pT_R = \frac{220 + 460}{407} = 1.67$$

$$pP_R = \frac{2750}{667} = 4.12$$

From Figure 15, $\quad Z = 0.848$

Number of lb moles, N, in 1000 standard

$$\text{cu ft of gas} = \frac{1000}{379}$$

From gas law equation,

$$V = \frac{NZRT}{P} = \frac{1000 \cdot 0.848 \cdot 10.73 \cdot (220 + 460)}{379 \cdot 2750}$$

Volume $= 5.93$ cu ft

Use of Pseudo-critical Constant-Gas Gravity Relations in Metering Gases

The calculation of gas meter factors at high pressures is a further application of data of Figures 15 and 17. The meter formula[15] may be expressed as

$$q = c\sqrt{\frac{1}{\gamma}}\sqrt{\frac{520}{T}}\sqrt{\frac{1}{Z}}\sqrt{Ph} \qquad (7)$$

where

q = standard cu ft of gas flowing per unit time
c = meter constant, dependent on orifice diameter, pipe diameter, etc.
γ = gas gravity (air = 1)
T = flowing temperature of gas, (°R)
Z = compressibility factor at flowing pressure and temperature
P = flowing pressure (psia)
h = orifice differential (inches of water)

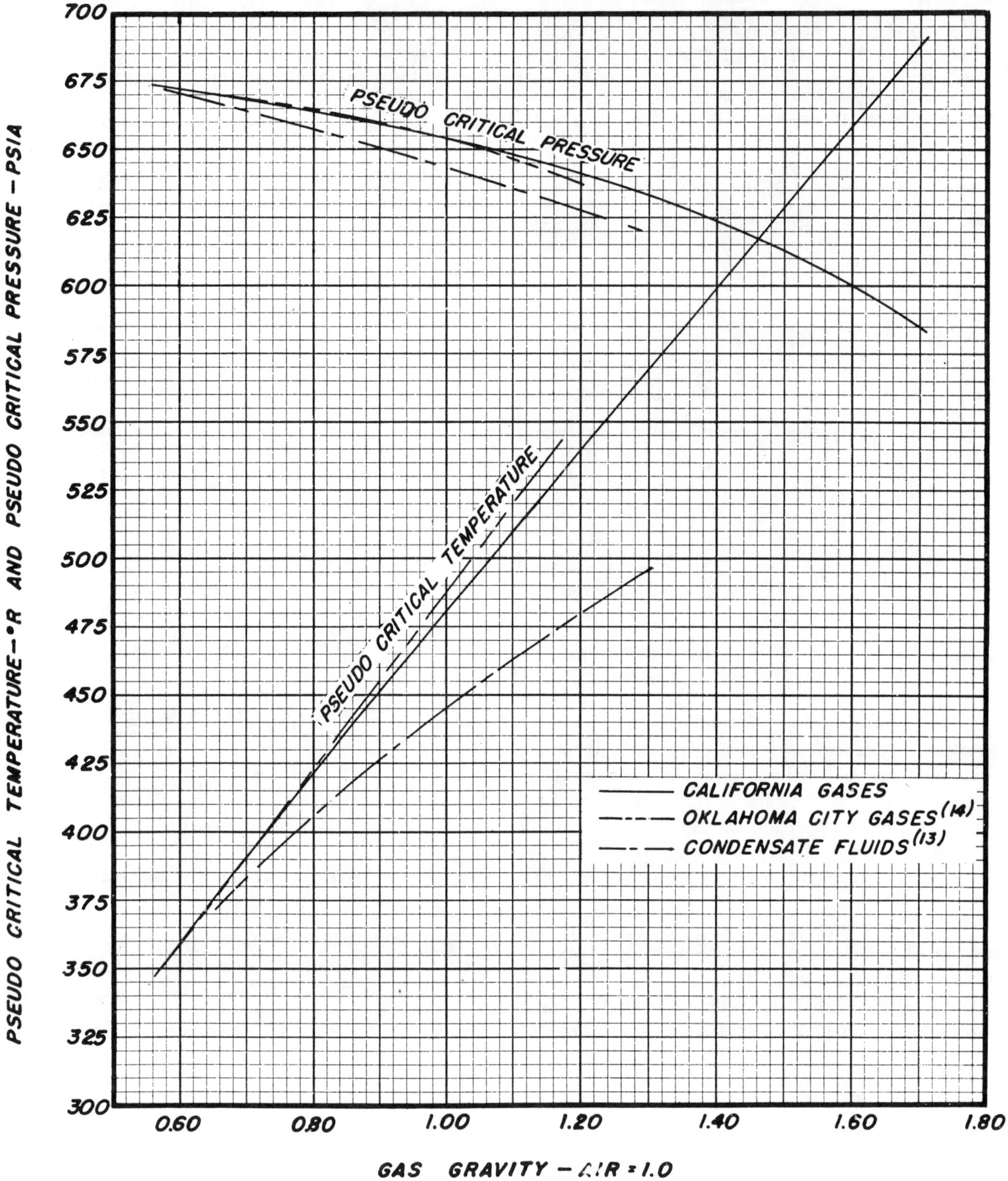

Figure 17—Psuedo-critical Constants of Gases and Condensate Well Fluids

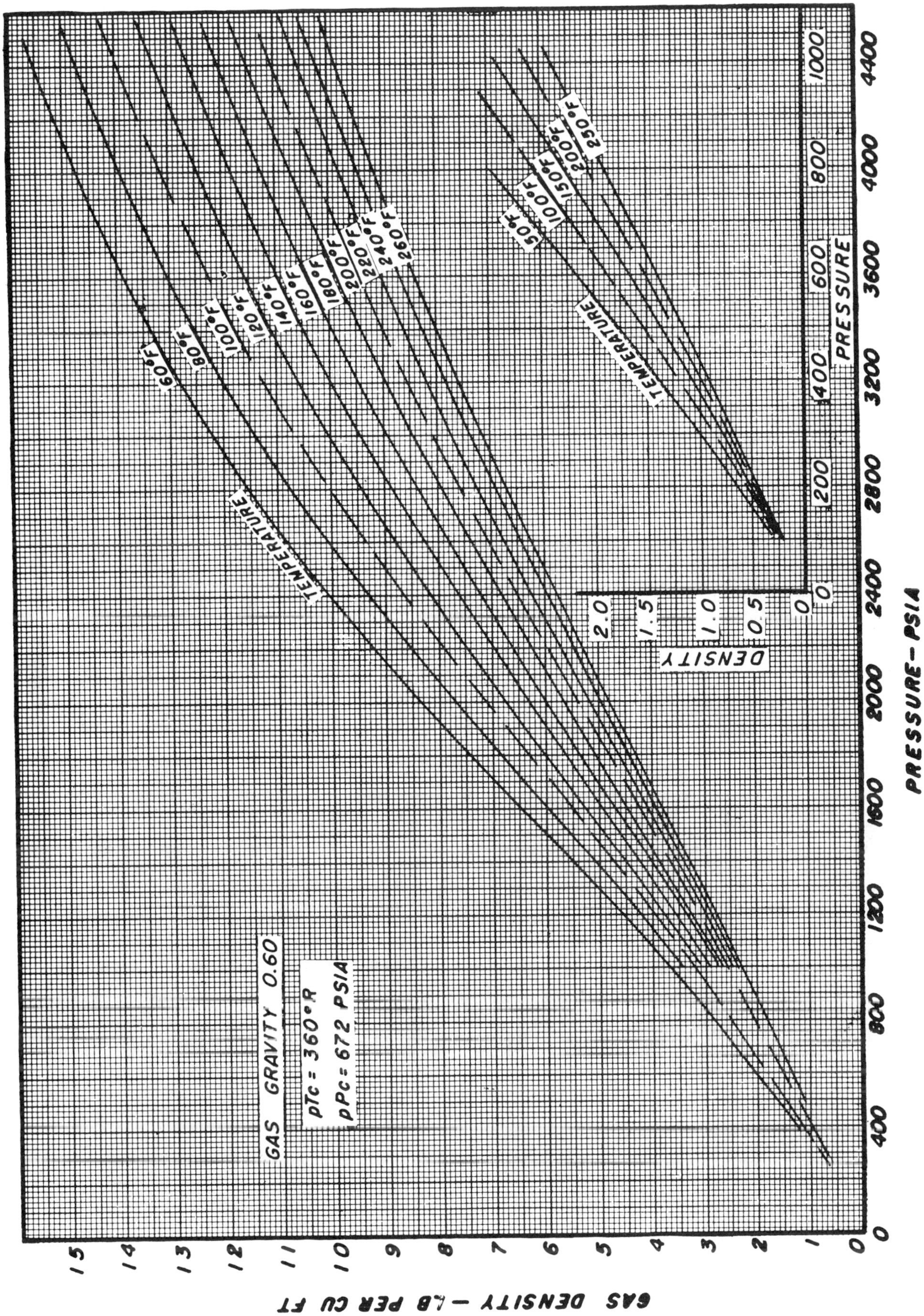

Figure 18—Pressure-Volume-Temperature Relations of 0.60 Gravity Natural Gas

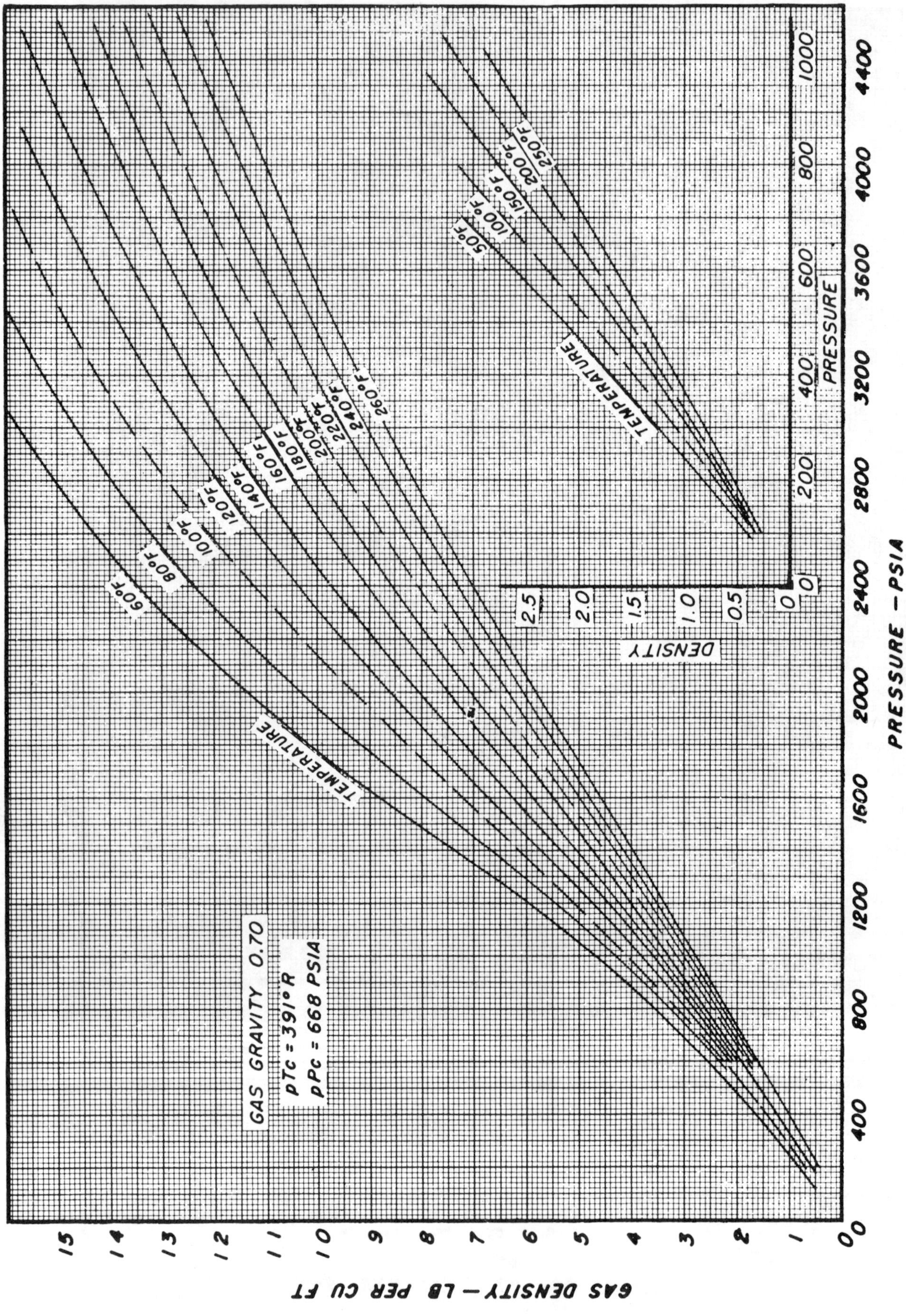

Figure 19—Pressure-Volume-Temperature Relations of 0.70 Gravity Natural Gas

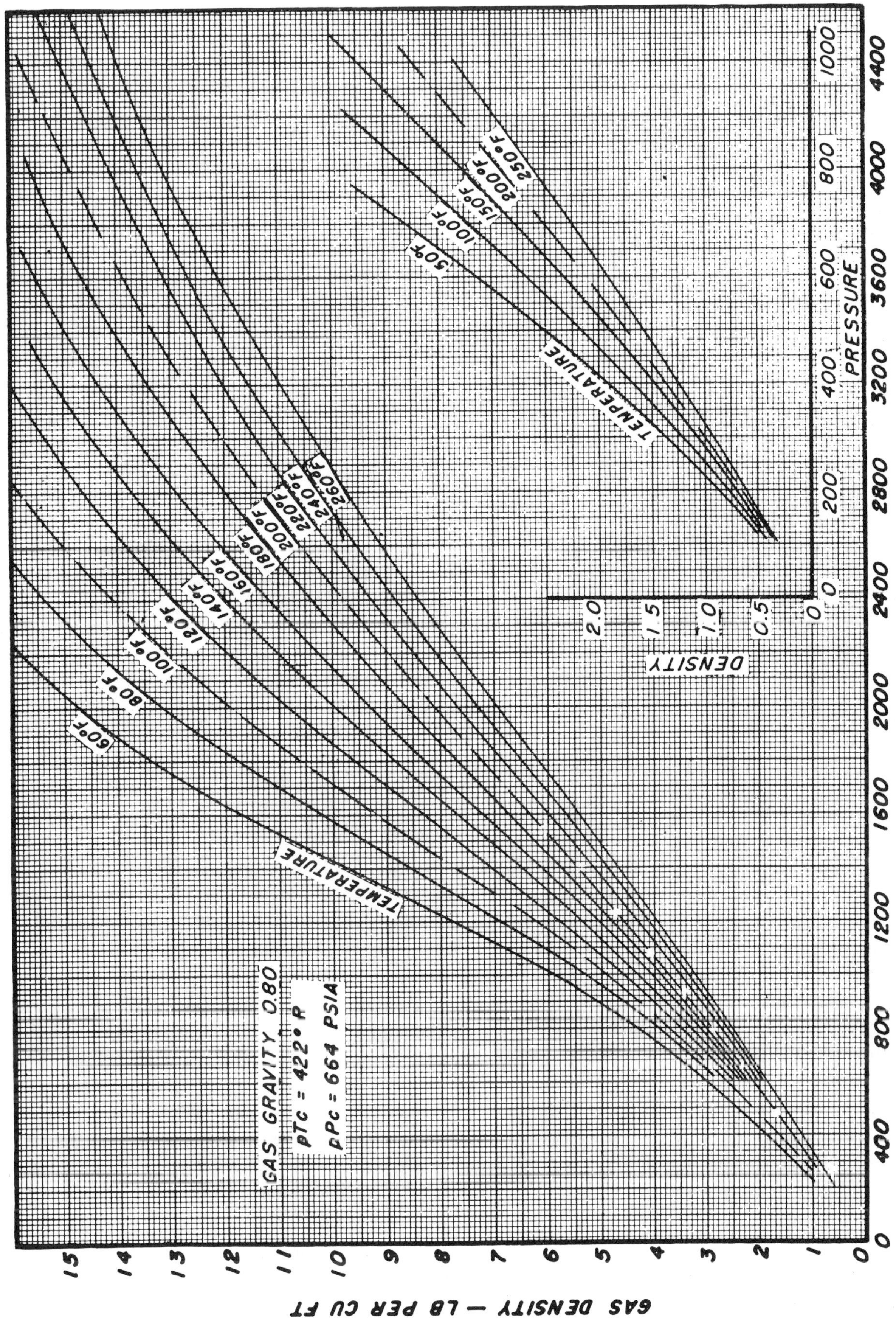

Figure 20—Pressure-Volume-Temperature Relations of 0 80 Gravity Natural Gas

The first three square-root terms may be combined as a single correction factor, $F_{\gamma TZ} = \sqrt{\frac{520}{\gamma TZ}}$ and the total factor plotted for any particular gas as a function of flowing temperature and pressure. An illustration of this is shown in Figure 21.

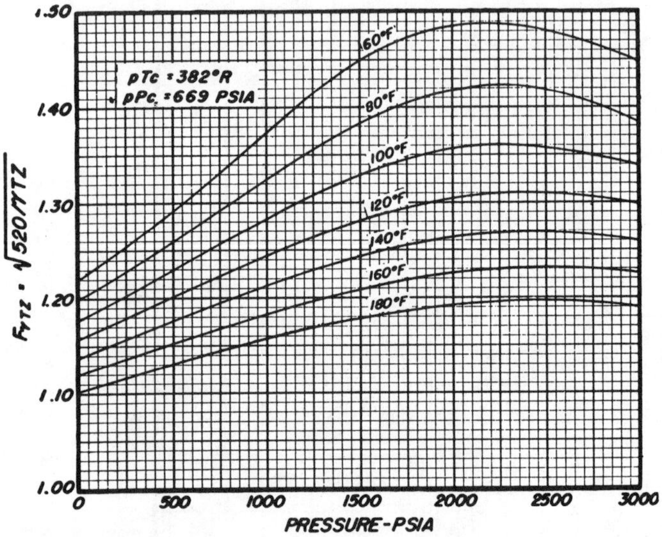

Figure 21—Meter Factors. Murphy-Coyote 0.67 Gravity Gas

Effect of Nonhydrocarbon Components on Compressibility Factors

Nitrogen and carbon dioxide are often present in appreciable concentrations in natural gases. As can be seen in Table 1, the critical constant-gravity relationship of these compounds does not fit in with that of the paraffinic hydrocarbons. It is to be expected, therefore, that correlations based solely on gas gravity can be in error when applied to gases containing large quantities of the nonhydrocarbons.

Eilerts, Mullens, and Carlson[16] have studied the compressibility of nitrogen and natural gas mixtures. They proposed that the compressibility of a nitrogen bearing natural gas be determined by separately evaluating the compressibility factors of the hydrocarbon portion and the nitrogen portion of the gas mixture and then obtaining an "additive" compressibility factor as follows:

$$Z_a = nZ_n + (1 - n)Z_h \tag{8}$$

where

Z_n = compressibility factor of the nitrogen in the gas sample

Z_h = compressibility factor of the mixture of hydrocarbons in the gas sample

n = mole fraction nitrogen in the sample

The computed value of the additive compressibility factor, Z_a, then has to be corrected to the actual compressibility factor, Z, by the equation:

$$Z = c \cdot Z_a \tag{9}$$

where c is an arbitrary factor to account for the fact that volumes of natural gas and nitrogen are not quite additive. Values of c, presented by the authors in chart form, indicate that the correction factor, c, is between 1.00 and 1.04 at temperatures above 70°F, and for nitrogen concentrations less than 20 per cent.

The above method is, of course, complicated by the requirement for c charts as well as compressibility charts both for the hydrocarbon gases and for nitrogen. Use of critical constants of nitrogen, 492 psia and 227°R, in the theorem of corresponding states yields a much easier way of calculating a compressibility factor. At nitrogen concentrations of less than 20 per cent, the compressibility factor of the entire gas evaluated by the theorem of corresponding states is usually in error less than four per cent.

Olds, Sage, and Lacey[17] investigated the behavior of carbon dioxide-methane mixtures. Reamer, Olds, Sage, and Lacey[18] determined compressibility factors for the ethane-carbon dioxide system. Application of the theorem of corresponding states will yield satisfactory results for natural gases having 2 per cent carbon dioxide or less. Olds, Sage, and Lacey indicated a possible error of 5 per cent in the compressibility factor at a concentration of 4 per cent carbon dioxide.

Residual Volume Methods

Residual volumes present the second or "additive" method of calculating PVT behavior of gases. The residual specific volume of an individual gas is defined by the equation:

$$\underset{\sim}{\overline{V}} = \overline{V}_{(perfect)} - \overline{V}_{(actual)} = \frac{RT}{MP} - \overline{V} \tag{10}$$

where

$\underset{\sim}{\overline{V}}$ = residual specific volume, cu ft per lb

$\overline{V}$ = specific volume, cu ft per lb

R = gas constant, per mole

M = molecular weight of the gas

T = temperature, °R

P = pressure, PSIA

Thus, the residual volume is the difference between the volume of a gas computed as an ideal gas and the actual gas volume. The residual becomes negative in the region of temperature and pressure where the actual gas occupies a volume larger than the corresponding volume of an ideal gas. (This is comparable to the region of Figure 15 wherein Z is greater than unity.)

The use of residual volumes becomes more complicated when a multicomponent gas is considered. Just as in the case of the compressibility factor method, it is nécessary to take into account the composition of the gas when calculating the deviation of the gas behavior from the ideal behavior. In calculating a compressibility factor, the weighting is accomplished by multiplying the mole fraction of each component by the critical constants of the component and summing the products. In the residual volume method the weighting is accomplished as shown by the following equations:

$$\bar{V} = \bar{V}_1 n_1 + \bar{V}_2 n_2 + \cdots \bar{V}_j n_j \qquad (11)$$

or as can be seen from Equation (10),

$$\bar{V} = \frac{RT}{M_a P} - \bar{V}_1 n_1 - \bar{V}_2 n_2 - \cdots - \bar{V}_j n_j \qquad (12)$$

where

M_a = average molecular weight of the whole gas

n = weight fraction of a component in the gas

$\bar{V}$ = residual partial specific volume of a component in the gas, cu ft per lb

The first term in Equation (12) represents the specific volume of the whole gas if it follows the ideal gas law, whereas the succeeding terms are the deviation from this idealized behavior of each of the components in the gas. In Equation (12), the residual partial volume of each component is a function of pressure, temperature, and the nature and amount of all the other components present in the gas.*

Sage and Lacey and their co-workers have studied systematically the behavior of binary systems composed of the lighter components found in natural gases. By 1939, residual partial specific volume data had been reported[19] on methane, ethane, propane, and n-butane at pressures to 3000 psia and at temperatures between 70°F and 250°F. Since then, partial volume data (from which residual partial volumes can be calculated) have been extended to 10,000 psia and 460°F.[21] Additional systems composed of methane-decane, and methane-n-butane-decane have been investigated.

Let us consider the use of residual partial specific volume data in computing the specific volume of a gas at an elevated pressure and temperature. In order to use the published residual data,[19] it must be assumed that the residual partial volume of the material that is heavier than isobutane can be represented by that of n-butane. If we let $(\bar{\bar{V}}_1)_{1:2}$, $(\bar{\bar{V}}_1)_{1:3}$, and $(\bar{\bar{V}}_1)_{1:4}$ represent the residual partial specific volume of methane in the methane-ethane, methane-propane, and methane-n-butane systems, and give weight to the individual methane values in direct proportion to the concentration of the heavier component in the three binary systems, the weighted residual partial volume of methane may be established from the following relation:

$$\bar{V}_1 = \frac{(\bar{V}_1)_{1:2}\, n_2 + (\bar{V}_1)_{1:3}\, n_3 + (\bar{V}_1)_{1:4}\, n_4}{1 - n_1} \qquad (13)$$

In absence of similar data to calculate weighted residual partial volumes of ethane, propane, and butane, it is necessary to assume that the value determined in combination with methane is applicable to any multicomponent system.

Substituting the value of $\bar{\bar{V}}_1$ in Equation (13) into the second term in Equation (12) gives the following equation:

$$\bar{V} = \frac{10.73\,T}{M_a P} - \left[\frac{n_1}{1-n_1}\right]\left[(\bar{V}_1)_{1:2}\, n_2 + (\bar{V}_1)_{1:3}\, n_3 + (\bar{V}_1)_{1:4}\, n_4\right] - \bar{V}_2 n_2 - \bar{V}_3 n_3 - \bar{V}_4 n_4 \qquad (14)$$

The calculation of the specific volume of 0.675 gravity gas at 3000 psia and 250°F, shown in Table 6, illustrates the use of Equation (14) and residual partial volume data. The length of the calculations would appear to be about equal to the length using compressibility factors. However. residual partial volumes are not interpolated easily between tabulated pressures and temperatures, so the usual calculation is actually more laborious than indicated here. When a problem calls for a

* The method of determining residual partial specific volumes and their mathematical significance has been given by Sage and Lacey.[10,20] The discussion will not be repeated here.

TABLE 6. EXAMPLE CALCULATION OF THE SPECIFIC VOLUME OF A GAS USING RESIDUAL VOLUME DATA

Pressure 3000 PSIA — Temperature 250° F

1	2	3	4	5	6	7	8
Component	Mole Fraction n	Molecular Weight of Component M	nM	Weight Fraction $\underline{n}$		$(\underset{\sim}{\bar{\bar{V}}}_1)_{1:-}$	$\underset{\sim}{\bar{\bar{V}}}$
C_1	0.8602	16.0	13.80	0.7128		—	—
C_2	0.0770	30.1	2.32	0.1201		0.00648	0.0352
C_3	0.0426	44.1	1.88	0.0973		0.0066	0.0320
i-C_4	0.0057	58.1	0.33	0.0171			
n-C_4	0.0087	58.1	0.51	0.0264			
i-C_5	0.0011	72.2	0.08	0.0041	Weight Fraction i-C_4+ = 0.0699	0.0037	0.0315
n-C_5	0.0014	72.2	0.10	0.0052			
C_6+	0.0033	100.	0.33	0.0177			
	1.0000		19.32	1.0000			

Values for $\underset{\sim}{\bar{\bar{V}}}_1$, $(\underset{\sim}{\bar{\bar{V}}}_1)_{1:2}$ etc. at 3000 PSIA and 250°F are from data tabulated in Reference 19

$$\text{Specific Volume, } \bar{V} = \frac{10.73T}{M_a P} - \left[\frac{\underline{n}_1}{1-\underline{n}_1}\right]\left[(\underset{\sim}{\bar{\bar{V}}}_1)_{1:2}\,\underline{n}_2 + (\underset{\sim}{\bar{\bar{V}}}_1)_{1:3}\,\underline{n}_3 + (\underset{\sim}{\bar{\bar{V}}}_1)_{1:4}\,\underline{n}_4\right] - \underset{\sim}{\bar{\bar{V}}}_2\underline{n}_2 - \underset{\sim}{\bar{\bar{V}}}_3\underline{n}_3 - \underset{\sim}{\bar{\bar{V}}}_4\underline{n}_4 =$$

$$\frac{10.73 \cdot 710}{19.32 \cdot 3000} - \frac{0.7143}{(1-0.7143)}\left[(0.00648)(0.1201) + (0.0066)(0.0973) + (0.0037)(0.0699)\right] - (0.0352)(0.1201)$$

$$-(0.0320)(0.0973) - (0.0315)(0.0683)$$

Specific Volume, $\bar{V}$ = 0.1168 cu ft per lb

solution of pressure, with the temperature, volume, and composition of the gas given, the calculations may become excessively complicated because the pressure must be roughly known in order to evaluate the residual partial volumes. In other words, it is necessary to resort to a trial and error calculation method of determining the pressure.

The accuracy of the residual volume method is at present no better than that of the compressibility factor method. For example, specific volumes of the gas, shown in Table 6 calculated at 3000 psia and 250°F by the three methods outlined in this chapter, gave the following agreements with the experimentally determined specific volume:

	Specific Volume (cu ft per lb)	Per Cent Error
Experimental	0.1217	—
By Figures 18 and 19	0.1203	−1.15
By pseudo-critical method	0.1197	−1.65
By residual volume method	0.1168	−4.40

Summary of Principal Points

The equation, $PV = NZRT$, and the two constants, 10.73 and 379, are the most important factors in making gas calculations. Remembering these three factors will allow one to solve any gas PVT problem.

When gravity with respect to air is the only known parameter of a gas, the PVT behavior of the gas is best obtained by use of Figures 18 to 20. At gas gravities greater than 0.8, the pseudo-critical-gas gravity relations shown in Figure 17 can be used to evaluate a compressibility factor at the desired pressure and temperature.

In gases containing only 1 or 2 per cent of heavy fractions, the true critical constants of the next higher compound should be used for the "plus" fraction. Thus, a hexanes plus fraction uses the critical constants of *n*-heptane.

CHAPTER 4

BEHAVIOR OF LIQUIDS

The PVT relation of liquids also is important in calculations pertaining to petroleum reservoir behavior. The isothermal compressibility of undersaturated liquid, the shrinkage in volume that a barrel of reservoir liquid undergoes while progressing from the reservoir to the lease tank, and the in-place density of the reservoir liquid are three examples that require a knowledge of hydrocarbon liquid behavior.

There are several methods of calculating the volume that a given mass of liquid occupies under an elevated pressure and temperature. When the composition of a reservoir liquid is known, its volume at any pressure and temperature can be calculated either by applying ideal solution principles or by using partial volume methods. When the composition of the liquid is unknown, but the properties and quantities of surface gas and oil derived from the reservoir liquid are known, empirical methods are available to calculate the liquid volume under reservoir conditions. This chapter will outline first the methods that can be used when composition of the liquid is known. Following this, two empirical methods will be presented that are based on surface gas-oil ratio measurements.

Method of Calculating Liquid Density from Composition Using Ideal Solution Principles

The behavior of mixtures of liquid hydrocarbons at atmospheric pressure and temperature approaches the behavior of ideal solutions. That is, a unit volume of one liquid, such as pentane, added to a unit volume of another liquid, such as heptane, will result in two volumes of mixture. This ideal solution behavior makes it possible to calculate the volume of a liquid mixture at 14.7 psia and 60°F from the composition of the liquid and the densities of the individual components. The atmospheric volume of the mixture can then be corrected for compressibility and thermal expansion to obtain its volume at an elevated pressure and temperature.

When applying the ideal solution method to reservoir liquid systems that contain large quantities of dissolved gases, it is apparent immediately that it is impossible for such systems to continue to be all liquid when brought to atmospheric pressure and temperature. However, this physical limitation does not impair the mathematical use of atmospheric liquid densities of methane and ethane in an ideal solution calculation, because the "pseudo-liquid density" of the system at atmospheric conditions is only a stepping stone to the evaluation of the density at an elevated pressure and temperature.

Standing and Katz[22] studied measured data on systems containing methane plus other compounds and ethane plus other compounds. By taking the experimentally determined system density at numerous elevated pressures and temperatures and correcting this density to 14.7 psia and 60°F, a so-called pseudo-liquid density of the system was calculated. Subtracting the volume and weight contributed by the "plus" compound from the pseudo-density left an "apparent liquid density" of the methane or ethane. In other words, the apparent liquid densities of methane and ethane are fictitious densities and are the values which, used in an ideal solution calculation with the true liquid densities of other compounds, give an atmospheric system density that can be corrected to elevated pressures and temperatures by suitable compressibility and thermal expansion factors.

For convenience, the apparent liquid densities are calculated at conditions of 14.7 psia and 60°F.

When the above procedure was carried out, the apparent liquid densities were found to vary with the density of the system as a whole, as shown in Figure 22. Direct use of the data in this form

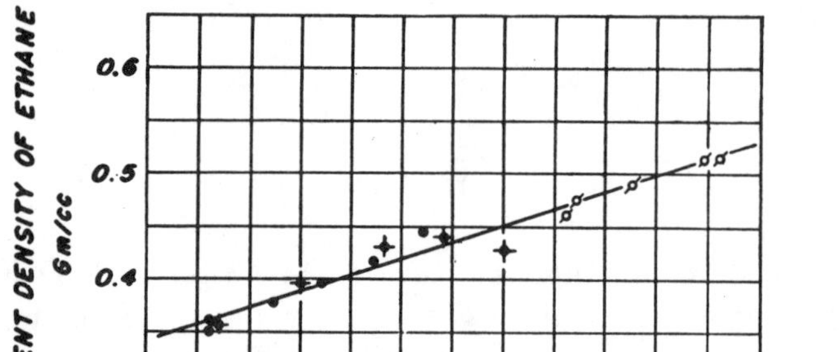

Figure 22—Relation of Apparent Density of Methane and Ethane to Density of the System (After Standing and Katz, Trans. AIME, 1942)

would require a trial-and-error solution; therefore, pseudo-liquid densities of the system have been converted into terms of the density of the heavier portion of the system and the weight per cent of the light component. The results of these conversions are presented in the form of the calculating chart shown in Figure 23. To use Figure 23, it is necessary to calculate the density of the propane plus fraction at 14.7 psia and 60°F, the weight per cent ethane in the ethane plus material, and the weight per cent methane in the entire system. From these three values the pseudo-liquid density of the system at 14.7 psia and 60°F is determined directly from the chart. This procedure is illustrated by the example problem given in Table 7.

Temperature-specific gravity corrections for liquid hydrocarbons having specific gravities in the range 0.60–1.0 are given in the National Standard Petroleum Oil Tables.[23] The NGAA Standard Correction Factors[24] cover a lower temperature range than the NSPO Tables, in addition to lower gravities. Sage, Lacey, and co-workers on API Project 37 have measured the effect of temperature and pressure on the density for a number of multicomponent liquids. A survey of these data indicates:

(1) The compressibility of hydrocarbon liquids at temperatures below 300°F can be assumed to be a function of the liquid density at 60°F. Minor differences exist in the liquid compressibility between systems of the same density, but the differences are not regular enough to take into account.

(2) The thermal expansion of hydrocarbon liquids is affected little by pressure. Thus, it is possible to use the temperature corrections listed in the National Standard Petroleum Oil Tables[23] at elevated pressures, as well as at atmospheric pressure.

An additive correction factor to the atmospheric density of a liquid for the effect of pressure has been presented by Standing and Katz.[22] The range of the factor was later extended by Hanson, Kuist, and Brown.[25] This correction factor, presented in Figure 24, makes it possible to correct the pseudo-liquid densities to an elevated pressure state. For example, the liquid, whose pseudo-liquid density was calculated to be 50.8 lb per cu ft at 14.7 psia and 60°F in Table 7, would have a density of $50.8 + 0.89 = 51.7$ lb per cu ft at 3280 psia and 60°F.

The correction for the effect of temperature change is presented in Figure 25. These factors were obtained from the data of National Petroleum Tables, the NGAA Standard Volume Correction Factors, and the data on low molecular weight liquids presented by Hanson, Kuist, and Brown. Applying the correction factors of Figure 25 to the example of Table 7 indicates that a correction

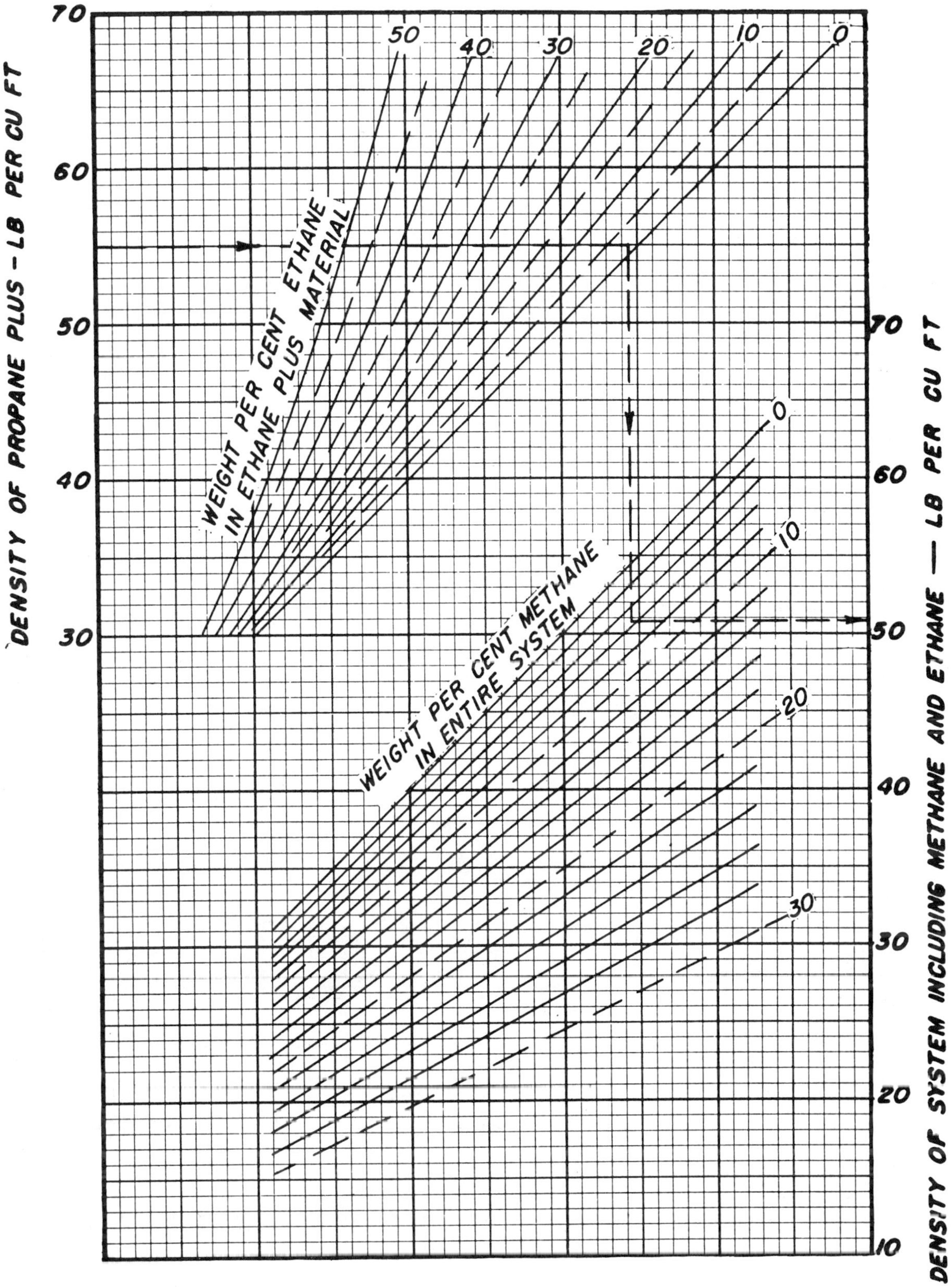

Figure 23—Pseudo Liquid Density of Systems Containing Methane and Ethane

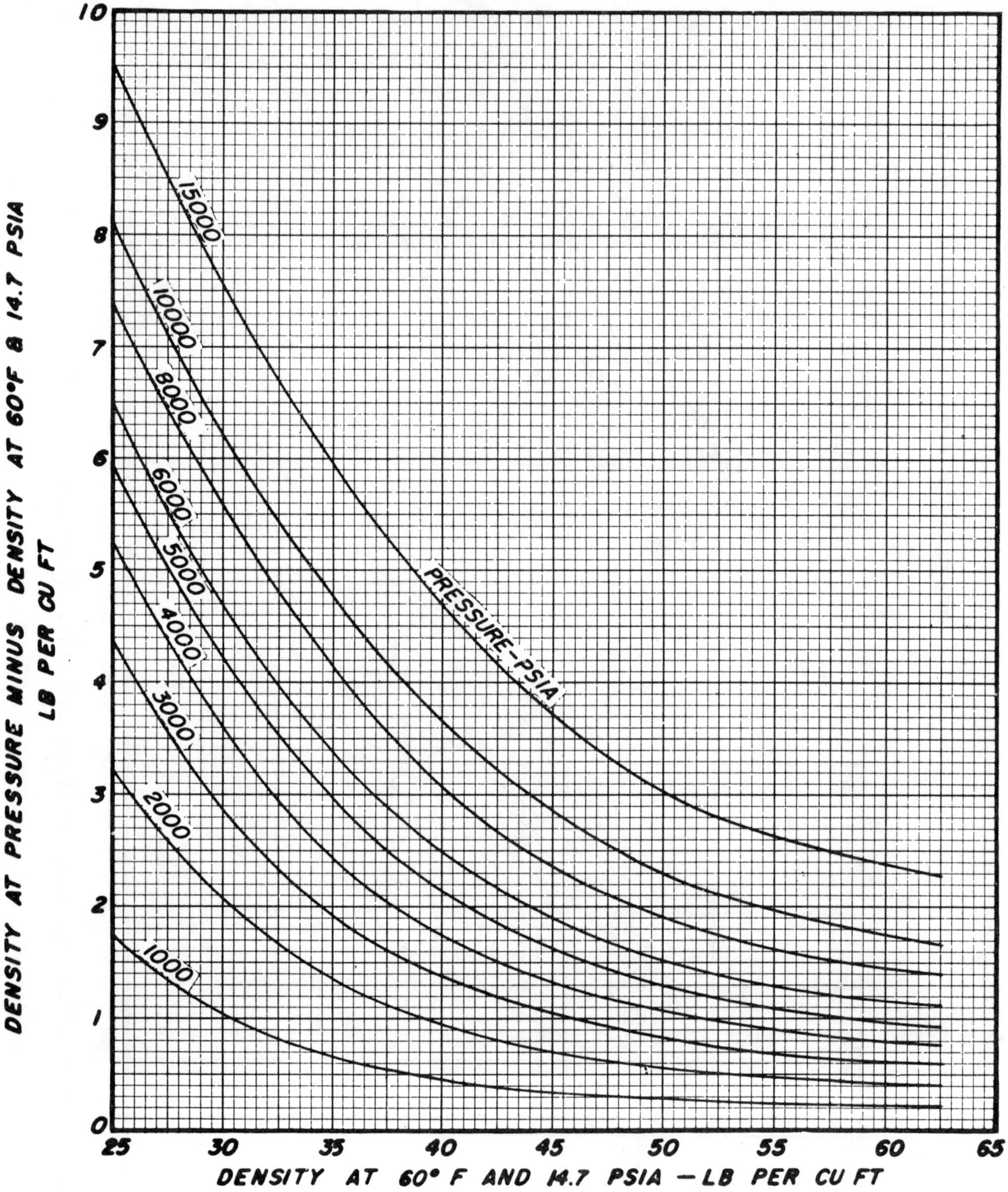

Figure 24—Density Correction for Compressibility of Liquids

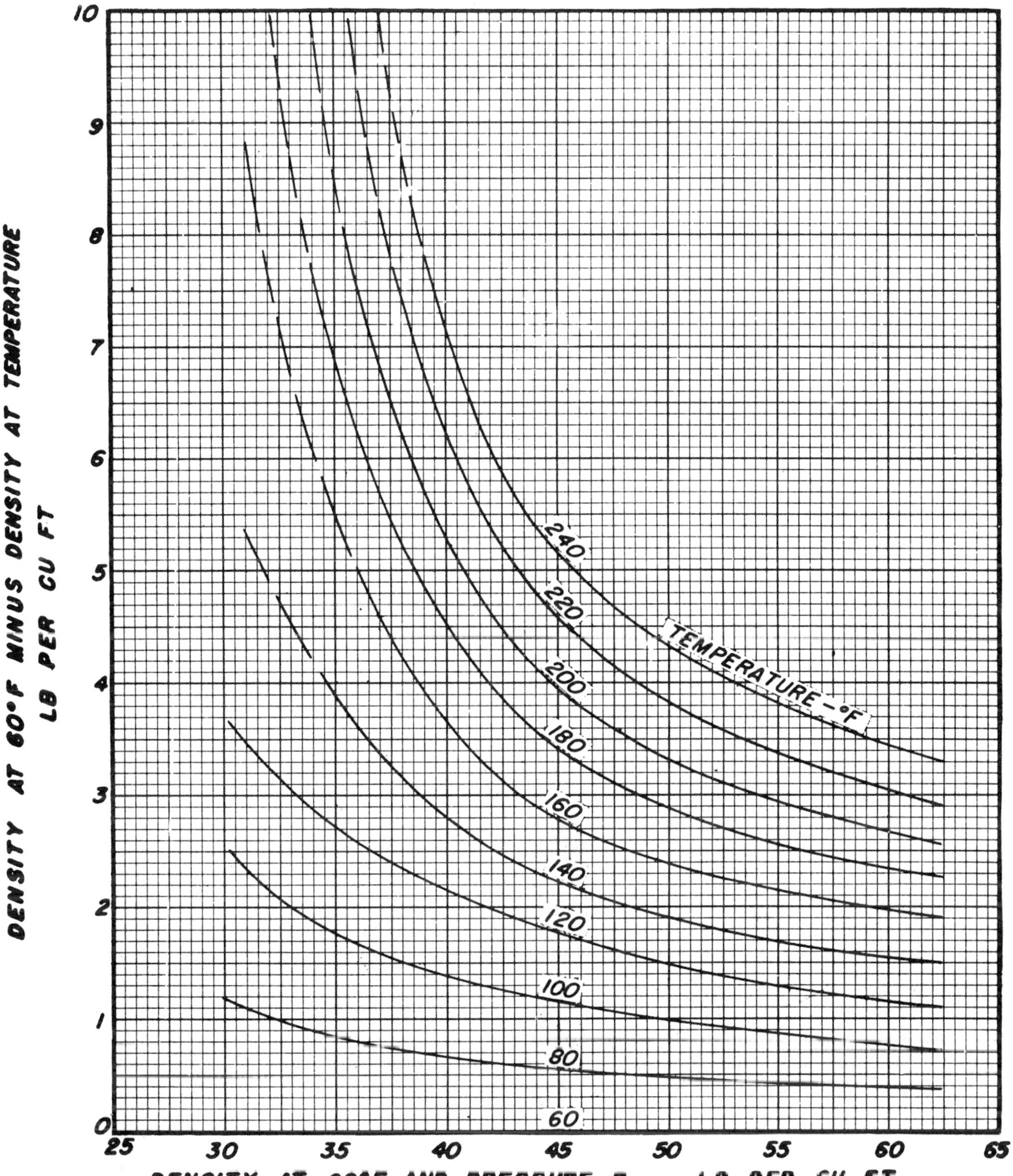

Figure 25—Density Correction for Thermal Expansion of Liquids

TABLE 7. CALCULATION OF LIQUID DENSITY USING IDEAL SOLUTION PRINCIPLES AND APPARENT METHANE AND ETHANE DENSITIES

Problem: To calculate the density of a crude oil-natural gas system at its bubble point of 3280 *psia at* 218°F, *given the composition of the system.*

Basis: One-pound mole.

1	2	3	4	5	6
Component	Mole Fraction n	Molecular Weight of Components M	Weight of Components lb n M	Liquid Density of Components at 60°F & 14.7 PSIA lb per cu ft ρ	Liquid Volume of Components at 60°F & 14.7 PSIA cu ft $\frac{nM}{\rho}$
C_1	0.4404	16.0	7.046		
C_2	0.0432	30.1	1.296		
C_3	0.0405	44.1	1.786	31.66	0.0564
C_4	0.0284	58.1	1.650	35.87*	0.0460
C_5	0.0174	72.2	1.256	39.16*	0.0321
C_6	0.0290	86.2	2.500	41.43	0.0603
C_7+	0.4011	287	115.1	56.6	2.032
	1.0000		130.69		2.227

*Arithmetic average of iso and normal values, Table 1

Density propane plus $= \frac{122.34}{2.227} = 54.93$ lb per cu ft

Weight per cent ethane in ethane plus $= \frac{1.296}{123.64} = 1.05$

Weight per cent methane in system $= \frac{7.05}{130.69} = 5.4$

From Figure 23, pseudo-liquid density = 50.8 lb per cu ft at 60°F and 14.7 psia
From Figure 24, pressure correction to 3280 psia = +0.89 lb per cu ft
Density at 3280 psia and 60°F = 50.8 + 0.89 = 51.7 lb per cu ft
From Figure 25, temperature correction to 218°F = −3.57 lb per cu ft
Density at 3280 psia and 218°F = 51.7 − 3.57 = 48.1 lb per cu ft

factor of −3.57 lb per cu ft should be applied to the 3280 psia-60°F density value to account for the thermal expansion of the compressed liquid in going from 60°F to 218°F. The density at 3280 psia and 218°F is, therefore, 51.7 − 3.57 = 48.1 lb per cu ft.

The method just outlined and illustrated by Table 7 usually will give densities less than 3 per cent in error. Standing and Katz[22] compared experimentally determined densities of fifteen liquids saturated with gas at pressures up to 8200 psi with calculated density values and found an arithmetic average error of only 1.16 per cent. The greatest error was four per cent. Checks with various data of Sage and Lacey have indicated comparable accuracy.

Method of Calculating Liquid Density from Composition and Partial Liquid Volumes

The specific volume of a liquid (reciprocal of density) at an elevated pressure and temperature state can be calculated by summing the products of the individual component weight fractions and their partial liquid volumes. This can be expressed mathematically as

$$\bar{V} = \bar{V}_1 n_1 + \bar{V}_2 n_2 + --- \bar{V}_j n_j \quad (15)$$

As in the gas calculations, the partial volumes of the individual components are a function of pressure, temperature, and the kind of material comprising the rest of the system. This third variable, in the case of hydrocarbon liquids found in petroleum reservoirs, is expressed by the average molecular weight of the other components in the system. Therefore, in using the partial volume method, it is necessary to evaluate both the weight fraction of every component in the system and the average molecular weight of all other components except the one under consideration. It can be appreciated that with three variables to consider, extensive cross-plotting of the data often is required.

TABLE 8. CALCULATION OF LIQUID DENSITY BY THE METHOD OF PARTIAL VOLUMES

Problem: *To calculate the density of a crude oil-natural gas system at its bubble point 3280 psia at 218°F, given the composition of the system.*

Basis: *One pound mole.*

1	2	3	4	5	6	7	8
Component	Mole Fraction n	Molecular Weight of Components M	Weight of Components lb n M	Weight Fraction $\underline{n}$	Average Molecular Weight of Other Components $\underset{\sim}{}$	Partial Volume cu ft per lb $\underset{\sim}{\bar{V}}$	Liquid Volume of Components at 3280 PSIA & 218°F $\underset{\sim}{\bar{V}}\,\underline{n}$
C_1	0.4404	16.0	7.046	0.0539	221	0.09	0.00485
C_2	0.0432	30.1	1.296	0.0099	135	0.0347	0.00034
C_3	0.0405	44.1	1.786	0.0137	134	0.0347	0.00048
C_4	0.0284	58.1	1.650	0.0126	133	0.0303	0.00038
C_5	0.0174	72.2	1.256	0.0096	131	0.0252	0.00024
C_6	0.0290	86.2	2.500	0.0191	23	0.0175	0.01575
C_7+	0.4011	287	115.1	0.8812			
	1.0000		130.69	1.0000			0.02204

Calculated density of liquid at 3280 PSIA and 218° F $= \frac{1}{0.02204} = 45.3$ lb per cu ft

Notes on Calculation Method

(1) Density of C_7+ fraction = 56.6 lb per cu ft at 14.7 PSIA and 60° F

(2) Partial volumes of methane through pentanes obtained from Sage, Hicks, and Lacey [26]. Methane value can be greatly in error because of limited data.

(3) Partial volume of C_6+ at 3280 and 218° F obtained by use of Figures 24 and 25 and the atmospheric density of C_6+ (56.1 lb per cu ft) to give a specific volume of 0.0186 cu ft per lb. Partial volume of 0.0175 then obtained from use of 0.0011 cu ft per lb residual volume and equation $\underset{\sim}{\bar{V}} = \bar{V} - \underset{\sim}{\bar{V}}$

An example calculation of the specific volume of a liquid at 3280 psia and 218°F is shown in Table 8. The partial liquid volume data of methane through pentane have been interpolated from the tabular data of Sage, Hicks, and Lacey.[26] The specific volume of the hexanes plus fraction at 14.7 psia and 60°F has been corrected by the data of Figures 24 and 25 to give the specific volume of this fraction at the elevated pressure and temperature. The partial volume of the hexanes plus was then obtained by subtracting a residual volume (tabulated by Sage, Hicks, and Lacey[26]) from the specific volume.

At the present time the accuracy of liquid density calculation made by the partial volume method is no better than that obtained by the empirical methods. Furthermore, the partial volume calculations are much more difficult to make. For these reasons, the partial volume method seldom is used. However, as the fund of partial volume data grows, the accuracy with which the density of liquids can be calculated will be increased. This probably is not true of the empirical methods.

Method of Calculating Liquid Density from Gas and Oil Properties and Apparent Gas Densities

The first section of this chapter has shown how methane and ethane assume different apparent liquid densities when in different composition systems. Katz,[27] in developing a method of predicting the shrinkage of crude oils, extended the apparent density concept from methane and ethane to cover gases in general. Katz's method has the advantage that it does not require the composition of the system, but can be used where only the gas-oil ratio, gas gravity, and tank oil gravity are known. The method is almost as accurate as the method using the composition of the system and normally will give the liquid densities to within 3 per cent.

The correlation found by Katz of the apparent density of dissolved gases as a function of gas gravity and tank oil gravity is given in Figure 26. An example of its use is worked out in Table 9, using gas gravities, tank oil gravity, and gas-oil ratio data to calculate the shrinkage, and a forma-

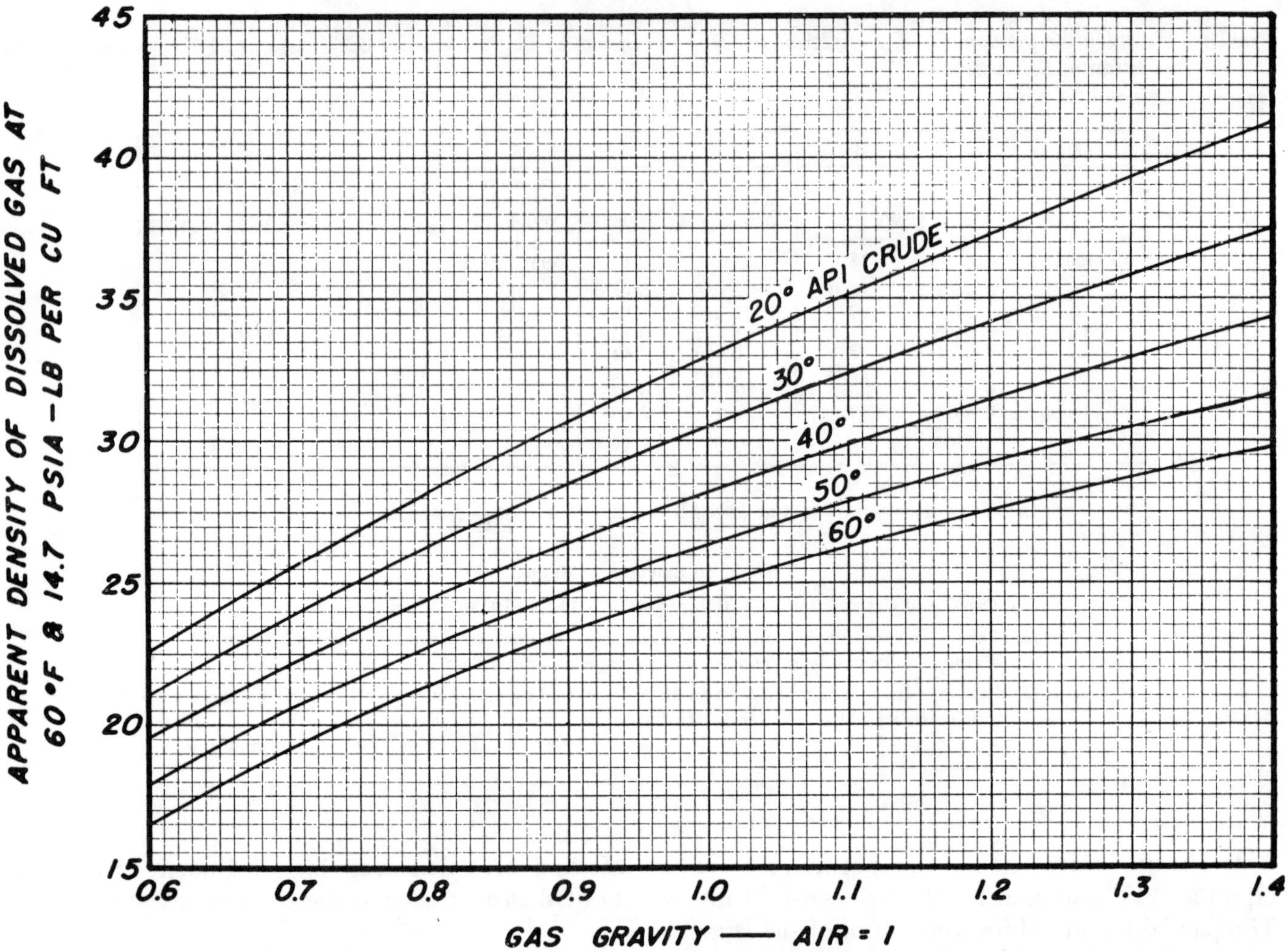

Figure 26—Apparent Liquid Densities of Natural Gases (After Katz, API Drilling and Production Practice, 1942)

tion volume factor of the liquid used in the previous illustrations. The calculated formation factor of 1.279 bbl per bbl of tank oil compares favorably with the experimentally determined value of 1.302 bbl per bbl of tank oil.

Method of Calculating Liquid Density from Generalized Data of California Crudes

A second empirical method that can be used to calculate a liquid density employs a correlation developed by Standing[7] for California crude oils and natural gases. This empirical correlation of the formation volume of a natural gas-crude oil system at its bubble point condition is shown in Figure 27 as a function of gas-oil ratio of the bubble point liquid, r_B; the dissolved gas gravity, γ_g; the tank oil specific gravity, δ_1; and temperature, t. The pressure is not used directly in this correlation, as it is an implicit function of the other variables. The formation volumes correlated in Figure 27 checked, with an error less than 0.5 per cent, 45 per cent of the 105 bubble point liquids comprising the correlation. The arithmetic average of all errors was 1.17 per cent. Although extensive checks have not been made with crudes from outside California, such crudes should not yield errors appreciably larger than those indicated above.

It will be noticed in the example shown in Table 9 that the calculation of the density and weight of a crude oil-natural gas system at its bubble point allows one to evaluate the formation volume at that condition. This relation can be used in reverse direction to calculate the density of a liquid from the weight of material in the system and the formation volume of the liquid at the desired pressure and temperature. An example of this method is shown in Table 10.

TABLE 9.' CALCULATION OF SHRINKAGE AND FORMATION VOLUME FACTOR OF A CRUDE OIL-NATURAL GAS MIXTURE USING APPARENT GAS DENSITY METHOD

Problem: To calculate the shrinkage that a reservoir liquid undergoes in being produced to the lease tank, given the surface production of gas and oil, the reservoir pressure and temperature, and assuming the liquid in the reservoir to be at its bubble point.

Given:

	Primary Trap	Secondary Trap	Tank	Total
Produced gas-oil ratio, cu ft per bbl tank oil	414	90	25	529
Produced gas gravity, air = 1	0.640	0.897	1.54	
Tank oil gravity	27.4° API = 0.891 sp gr			

Solution: Basis: One barrel tank oil

$$\text{Average produced gas gravity} = \frac{414 \cdot 0.640 + 90 \cdot 0.897 + 25 \cdot 1.54}{414 + 90 + 25} = \frac{384.2}{529} = 0.726$$

$$\text{Weight dissolved gas} = \frac{529}{379} \cdot 0.726 \cdot 28.97 = 29.3 \text{ lb}$$

Weight tank oil = 0.891 · 5.615 · 62.4 = 312 lb

From Figure 26,

Apparent density of dissolved gas = 24.9 lb per cu ft at 14.7 psia and 60°F

$$\text{Pseudo-liquid volume of dissolved gas} = \frac{29.3}{24.9} = 1.177 \text{ cu ft}$$

	Volume	Weight
Dissolved gas	1.177 cu ft	29.3 lb
Tank oil	5.615 cu ft	312 lb
	6.792 cu ft	341.3 lb

$$\text{Pseudo-liquid density} = \frac{341.3}{6.792} = 50.25 \text{ lb per cu ft at 14.7 psia and 60°F}$$

From Figure 24, pressure correction to 3280 psia = +0.90 lb per cu ft

Density at 3280 psia and 60°F = 50.25 + 0.90 = 51.15 lb per cu ft

From Figure 25, temperature correction to 218°F = −3.63 lb per cu ft

Density at 3280 psia and 218°F = 51.15 − 3.63 = 47.52 lb per cu ft

$$\text{Volume occupied at 3280 psia and 218°F (bubble point)} = \frac{341.3}{47.52} = 7.180 \text{ cu ft}$$

$$\text{Shrinkage} = \frac{7.180 - 5.615}{5.615} = 27.9 \text{ per cent, or}$$

$$\text{Formation volume} = \frac{7.180}{5.615} = 1.279 \text{ bbl per bbl tank oil}$$

TABLE 10. CALCULATION OF THE LIQUID DENSITY OF A CRUDE OIL-NATURAL GAS SYSTEM USING CORRELATION OF CALIFORNIA CRUDES

Problem: To compute the density of a saturated liquid from gas solubility and tank oil gravity data.

Given:

	Primary Trap	Secondary Trap	Tank	Total
Produced gas-oil ratio, cu ft per bbl tank oil	414	90	25	529
Produced gas gravity, air = 1	0.640	0.897	1.54	
Tank oil gravity	27.4° API = 0.891 sp gr			

Bubble point pressure 3280 psia at 218°F

TABLE 10—(*Continued*)

Solution: Basis: One barrel tank oil

Average produced gas gravity

$$= \frac{414 \cdot 0.640 + 90 \cdot 0.897 + 25 \cdot 1.54}{414 + 90 + 25} = \frac{384.2}{529} = 0.726$$

$$\text{Weight dissolved gas} = \frac{529}{379} \cdot 0.726 \cdot 28.97 = 29.3 \text{ lb}$$

Weight tank oil = 0.891 · 5.614 · 62.4 = 312 lb

Total 341.3 lb

$$r_R \left(\frac{\gamma_g}{\delta_1}\right)^{0.5} + 1.25t = 529 \left(\frac{0.726}{0.891}\right)^{0.5} + 1.25 \cdot 218 = 749.7$$

From Figure 27, formation volume = 1.32 bbl per bbl tank oil

$$\text{Density} = \frac{341.3}{1.32 \cdot 5.615} = 46.0 \text{ lb per cu ft at 3280 psia and 218°F}$$

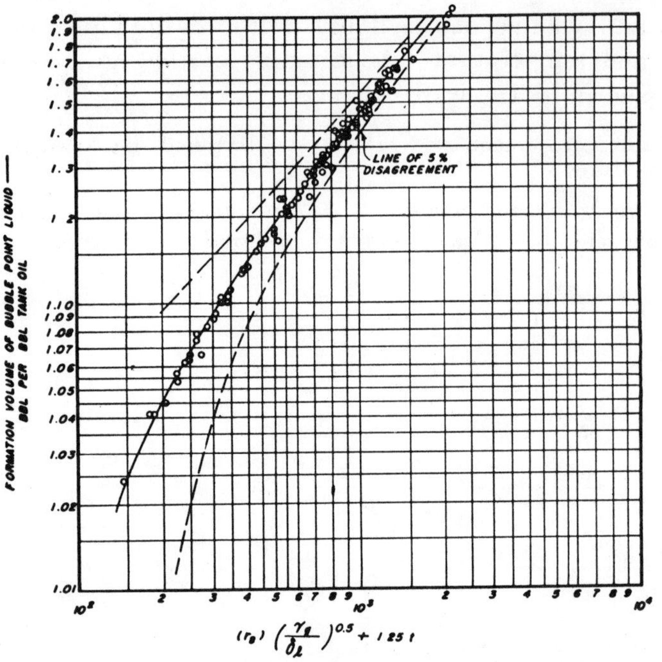

Figure 27—Formation Volume of Bubble Point Liquids from Gas-Oil Ratio, Dissolved Gas Gravity, Tank Oil Specific Gravity, and Temperature

Summary of Principal Points

Four methods have been outlined of calculating the density of a liquid at its bubble point condition of 3280 psia and 218°F. For the single liquid used in the illustration, the following answers were obtained.

	Density at 3280 psia and 218°F lb per cu ft	Per Cent Error
Experimental	46.4	.....
Ideal solution method	48.1	+3.7
Partial volume method	45.3	−2.4
Apparent gas density method	47.5	+2.4
California crude correlations	46.0	+0.9

When a liquid contains only minor amounts of methane and ethane (for example, a primary trap liquid or a tank oil), "average" values of 20.6 lb per cu ft for methane (0.33 gm per cc) and 29.0 lb per cu ft for ethane (0.45 gm per cc) can be used to represent these components in the ideal solution method. This shortens the calculation by not requiring the evaluation of propane plus properties and the weight per cent methane and ethane in the system.

Use an "average" apparent liquid density of 29.9 lb per cu ft at 14.7 psia and 60 °F for dissolved nitrogen.

The information in Figures 24, 25, and 27 can be used to determine the compressibility and thermal expansion of reservoir liquids when direct measurements are lacking. For example, using the method outlined in Table 10, a density of 46.0 lb per cu ft was obtained for the liquid at its bubble point. By applying the data of Figure 24, the bubble point liquid is calculated to have an undersaturated density of 46.62 lb per cu ft at 5000 psia and 218°F. The average compressibility of the liquid over the indicated pressure interval is

$$46.31\left(\frac{1/46.62 - 1/46.0}{5000 - 3280}\right) = -7.8(10^{-6})$$

volume per volume per psi.

CHAPTER 5

VAPOR-LIQUID EQUILIBRIA

It will be recalled from the phase behavior discussions presented in Chapter 1 that an area bounded by the bubble point and dew point curves on the pressure-temperature diagram of a multicomponent system defines the conditions for vapor and liquid to exist in equilibrium. At different points within the limits of the phase boundary, the quantity and composition of both phases vary. The purpose of this chapter is to present methods of calculating the quantity and composition of equilibrium vapor and liquid existing at any specified temperature and pressure. These methods have wide applicability to calculations of reservoir fluid behavior and to calculations concerned with processing reservoir fluids at the surface for their natural gas and heavier oil components.

The distribution of a component of a system between vapor and liquid is expressed by the equilibrium constant, K. The equilibrium constant is defined as the ratio of the mole fraction of the component in the vapor phase, y, to the mole fraction of the component in the liquid phase, x; mathematically, then

$$K = \frac{y}{x} \tag{16}$$

where
K = equilibrium constant of a component
y = mole fraction of a component on the vapor phase
x = mole fraction of a component in the liquid phase

The equilibrium constants of the various components of petroleum are functions of pressure, temperature, and the over-all composition of the system. At low pressure the effect of the system composition is small; but above 1000 psia the over-all composition of the system considerably affects the equilibrium constant. The word "constant" is a misnomer, but is used widely in the petroleum industry. Muskat[28] has proposed the term "equilibrium ratio" to indicate that the ratio is not solely a function of pressure and temperature. Although his term is more meaningful than "equilibrium constant," the old term will be used here.

Equilibrium constants can be evaluated by three methods: from Raoult's and Dalton's laws, from data on vapor and liquid fugacities of pure compounds, and by direct analysis of the vapor and liquid in equilibrium at any specified pressure and temperature. After the three methods of evaluating equilibrium constants are discussed in detail, several methods of using them will be explained.

Raoult's and Dalton's Laws

For ideal solutions at equilibrium, Raoult's law and Dalton's law may be combined to calculate concentration of each component in the vapor and liquid phases. Raoult's law states that the partial pressure of a component is equal to the mole fraction of the component in the liquid phase times the vapor pressure of the pure component. Dalton's law states that the partial pressure of a component in a vapor is equal to the total pressure exerted by the vapor times the mole fraction of the component in the vapor. Raoult's and Dalton's laws may be expressed as follows:

$$\underset{\sim}{pp} = x \cdot p \quad \text{(Raoult's Law)} \tag{17}$$

$$\underset{\sim}{pp} = y \cdot P \quad \text{(Dalton's Law)} \tag{18}$$

where
$\underset{\sim}{pp}$ = partial pressure of a component in the vapor phase
x = mole fraction of a component in the liquid phase
y = mole fraction of a component in the vapor phase
p = vapor pressure of the pure component at the temperature of the system
P = total pressure of the system

Combining the equations of these two laws, we have

$$x \cdot p = y \cdot P \tag{19}$$

which can be arranged to give

$$\frac{p}{P} = \frac{y}{x} = K \tag{20}$$

K is the equilibrium constant at the particular temperature and total pressure.

Two factors that restrict the use of equilibrium constants calculated by Raoult's and Dalton's laws are immediately apparent.

(1) A pure compound cannot have a vapor pressure at temperatures above its critical temperature. Thus, Raoult's law is limited to temperatures less than the critical temperature of all compounds in the system. This means, for example, that an equilibrium constant for methane cannot be evaluated by Raoult's law at temperatures above −116°F.

(2) Dalton's partial pressure law assumes that each component in the vapor behaves as an ideal gas. For practical purposes, the ideal gas assumption places an upper limit of pressure of 50 to 100 psia.

Equilibrium constants calculated from Equation (20) take the form of straight lines of unit slope on log K vs log P coordinates, and have values of unity at the pressure corresponding to the vapor pressure of the component at the temperature under consideration. For example, the vapor pressure of propane at 120°F is 240 psia (see Figure 3) so that the equilibrium constant for propane at 120°F takes the form shown in Figure 28. Constants calculated from Raoult's law are no longer important because ideal equilibrium constants and empirical constants far more accurately describe multicomponent system behavior.

Ideal Equilibrium Constants

A better approximation of actual equilibrium constants is obtained by defining the constants in terms of the fugacities of the compounds.[29,30] The fugacity may be looked upon as a vapor pressure modified to represent correctly the escaping tendency of the molecules from one phase into the other.* By using fugacities to define equilibrium constants, the departure of the vapor from ideal gas behavior and the effect of the total system pressure on the vapor pressure of the components are taken into account.

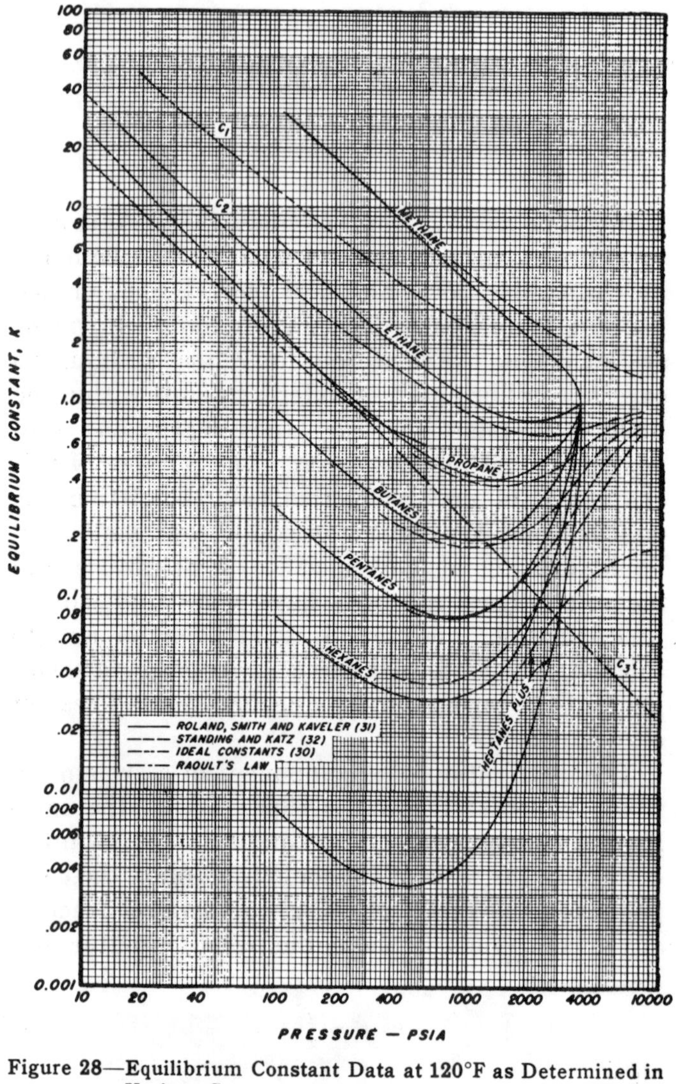

Figure 28—Equilibrium Constant Data at 120°F as Determined in Various Systems

In an ideal solution the fugacity of any compound in the system is equal to fugacity of the compound in the pure state times its mole fraction. Thus, if f'_{kl} and f'_{kg} are the fugacities of a compound in the liquid and vapor phases, and f_{kl} and f_{kg} are fugacities in the pure states,

$$f'_{kl} = f_{kl}x; \qquad f'_{kg} = f_{kg}y \tag{21}$$

* Refer to Sage and Lacey's "Volumetric and Phase Behavior of Hydrocarbons" (Reference 8) for more details regarding use of fugacities in determining ideal equilibrium constants.

At equilibrium conditions the fugacities of the compound are equal in the liquid and vapor phase, so that

$$\frac{y}{x} = \frac{f_{kl}}{f_{kg}} = K_{(\text{ideal})} \qquad (22)$$

Ideal constants are limited by one important fact. At a given temperature a compound can exist only as a single phase at any pressure other than the vapor pressure, so that the fugacity of either the vapor or liquid (depending upon whether the pressure is greater or less than the vapor pressure of the compound) must be obtained by extrapolation of the fugacity-pressure relation. As the extrapolation is somewhat arbitrary, the farther the pressure is from the vapor pressure, the more uncertain the equilibrium constant becomes. Usually the ideal equilibrium constants are considered undependable at pressures more than twice the vapor pressure of the compound. For this reason, they have limited usefulness for calculations dealing with petroleum reservoir conditions.

Empirical Equilibrium Constants

The third method of evaluating equilibrium constants is to determine experimentally the equilibrium compositions of vapor and liquid phases at various temperatures and pressures. The equilibrium constant of each component can then be evaluated by calculating the ratio of the mole fractions in the vapor and liquid phases.

When done correctly, this procedure results in a set of equilibrium constants representing a specific system. Figure 25 shows two sets of such constants at 120°F. The data of Roland, Smith, and Kaveler[31] were determined for a Gulf Coast condensate system containing roughly 70 mole per cent of methane and 21 per cent of hexanes plus material. The system studied by Standing and Katz[32] also was a condensate system. It contained about 83 mole per cent of methane and 6 mole per cent of hexanes plus. The constants of both systems are in good agreement at 1000 psia and, on the basis of other data, are expected to continue in agreement at lower pressures. However, the wide differences in the constants of the two systems at pressures approaching 3000 psia reflect the effect of system composition. Unfortunately, too few systems have been studied to define the manner in which the composition influences the values of the constants.

The shape of the log K-log P curves shown in Figure 28 is characteristic of all multicomponent system curves. At low pressures the slope of the curves is near minus one. As pressure increases, all the equilibrium constants decrease and, with the exception of methane, all pass through a minimum value. The constants tend to converge to unity at some high pressure.

The apparent convergence of the equilibrium constants to unity at some pressure is often misinterpreted to mean that the system is near its critical state. Unless the temperature at which the constants are determined is the critical temperature of the system, it is impossible for the equilibrium constants actually to converge, because it is only at the critical pressure and temperature of the system that the phase compositions are the same. At any temperature other than the critical temperature, the equilibrium constants become meaningless after the bubble point or dew point pressure of the system is reached. Nevertheless, isothermal log K-log P curves are often extrapolated to pressure regions wherein two phases cannot possibly exist. Quite often the curves have an appearance of converging. Therefore, in order to forestall confusion between actual behavior of systems at their critical state and their behavior postulated from extrapolated equilibrium constant data, it should be recognized that in the great majority of instances equilibrium constants converge to an "apparent convergence pressure," and that this pressure actually may be quite different from the critical pressure of the system.

The equilibrium constants of the plus fraction often behave in a manner different from the other components of a system. This is because the plus fraction in itself is a mixture of compounds, and both its vapor and liquid phase composition can vary with pressure.

In Figure 29 are plotted the specific gravities and molecular weights of the heptanes plus fraction[32] whose equilibrium constants are shown in Figure 25. At 4000 psia, for example, the equilibrium was between 130 molecular weight material in the vapor phase and 255 molecular material in the liquid phase. Increasing the pressure to 6000 psia caused the

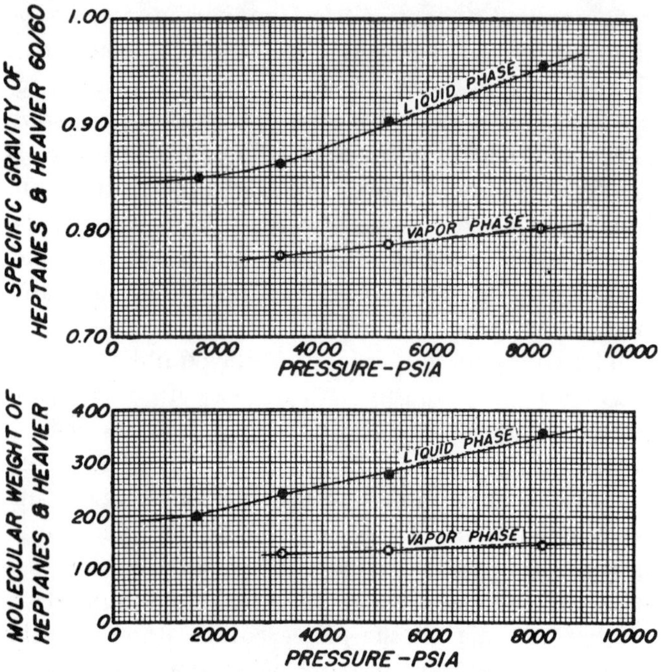

Figure 29—Effect of Pressure on Specific Gravity and Molecular Weight of Heptanes and Heavier (After Standing and Katz, Trans. AIME, 1944)

lighter portions of the liquid heptanes plus material to transfer to the vapor, raising the molecular weight of the vapor heptanes plus from 130 to 138. In so doing, the heptanes plus remaining in the liquid phase increased from 255 to 300. Therefore, it is apparent that at 6000 psia the heptanes plus equilibrium constant is for a material completely different from the material in equilibrium at 4000 psia. As a result, the log K-log P curve of the heptanes plus fraction has a shape different from that of the pure compounds in the system.

At present there are too few data with which to study the pressure behavior of the plus fraction. The explanation offered above is logical and, on the basis of the explanation, it is expected that the effect would be more noticeable when approaching dew point conditions than when near bubble points.

Construction of Log K-Log P Curves

At pressures less than 1000 psia, the influence of system composition on the equilibrium constant values usually is small. Calculations concerned with dissolved gas systems normally are made by using equilibrium constants determined by Katz and Hachmuth[33] for mixtures of natural gas and crude oil from the Oklahoma City field. The equilibrium constant data of Roland, Smith, and Kaveler[31] are used for condensate system calculations. Often these data are used also for dissolved gas and condensate system calculations at pressures above 1000 psia.

Above 3000 psia, the data of Katz and Hachmuth and of Roland, Smith, and Kaveler become less useful because of the apparent convergence of the constants near 4000 psia. It therefore becomes necessary to employ several empirical factors, in addition to pressure and temperature, to evaluate the constants in and above this pressure region.

Gilliland and Scheeline[34] and Brown and White[35] have shown how the apparent convergence pressure and the general shape of the log K-log P isotherms can be used to estimate equilibrium constants when experimental data are lacking. Referring to the general-shape curves shown in Figure 30, it can be seen that the components of a system can be divided

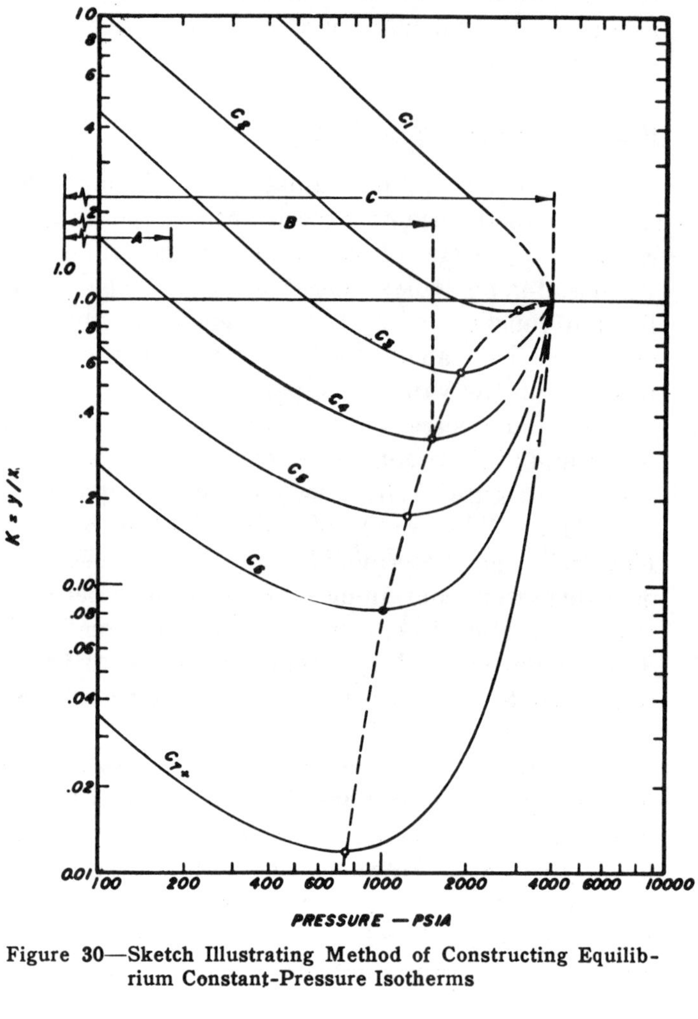

Figure 30—Sketch Illustrating Method of Constructing Equilibrium Constant-Pressure Isotherms

into two groups. The first group, referred to as more volatile components, contains those components whose equilibrium constants are never less than unity. The equilibrium constants of the less volatile components comprising the second group have minimum values, as well as two values of unity, one at the apparent convergence pressure and the other at the vapor pressure of the component. Thus, in the example shown in Figure 30, the equilibrium constant of butane has a minimum value of 0.32 at about 1500 psia and unity values at 4000 psia and 180 psia.

Brown and White found that by using the ratio of the apparent convergence pressure to the vapor pressure as one correlating factor, a correlation could be obtained with the ratio of the pressure for minimum K to the apparent convergence pressure. In other words, the ratio of distance C, Figure 30, to distance A can be used to obtain an estimate of the ratio of the distance B to distance C. Referring to the correlation of these factors shown in Figure 31, the value of the ordinate of the example cited above is $4000/180 = 22.2$. From the correlation, the ratio of the minimum K pressure to the apparent convergence pressure is found to be 0.28. Therefore the pressure at which the minimum butane K occurs is $4000 \cdot 0.28 = 1120$ psia.

The ratio of the apparent convergence pressure to the vapor pressure of a component also has been correlated to give values of the minimum constant. Referring to Figure 32, the minimum value of the butane constant in the example is 0.37. The equilibrium constants of other less volatile components can be obtained in like manner by using the apparent convergence pressure of the system and the vapor pressures of the components obtained from Figure 3.

The equilibrium constants of the more volatile components cannot be correlated by the methods outlined above. Sage, Hicks, and Lacey[36] have presented methane equilibrium constants as a function of the average molecular weight of the less volatile components in the system. Their correlations can be used at pressures below 1500 psi to define the general course of the methane curve. At pressures approaching the apparent convergence pressure, it usually is necessary to sketch in the curve.

One major difficulty hinders the use of the correlations presented in Figures 31 and 32 to determine K's for systems occurring in petroleum reservoirs. At the present time the apparent convergence pressure of a system cannot be evaluated from the composition of the system. Consequently, while Figures 31 and 32 will assist one to draw log K-log P isotherms, there is no direct way of knowing whether valid curves have been constructed for the particular system being studied. If PVT data are available from which to predict bubble point or dew point pressures, the selection of an apparent convergence pressure is simplified because the apparent convergence pressure must be higher than a bubble point or dew point pressure. The question then is: When the composition of a system is known, what is the procedure for determining constants for use at high pressures? The following is the author's method which, admittedly, leans heavily on the construction of curves "by eye." The absence of other published methods or correlations almost forces the use of this or some comparable method.

(1) Katz and Hackmuth's constants are accepted for use with dissolved gas systems. Roland, Smith, and Kaveler's constants are applied to condensate systems. When PVT data are available to define the bubble point or dew point pressure of the system, the constants are modified so that the calculated bubble point (or dew point) pressure is in agreement with the experimentally determined bubble point pressure. The modifications involves the following steps:

(a) The published data at the desired temperature are plotted on log K-log P coordinates. The bubble point pressure is then calculated, using the unmodified data. This procedure establishes a set of constants that will give a calculated bubble point, although the calculated bubble point pressure and the experimental bubble point pressure may not agree. By using the unmodified data, the experimentally determined relationship of the constants to each other is maintained.

(b) The set of K's that satisfactorily calculate a bubble point condition are then

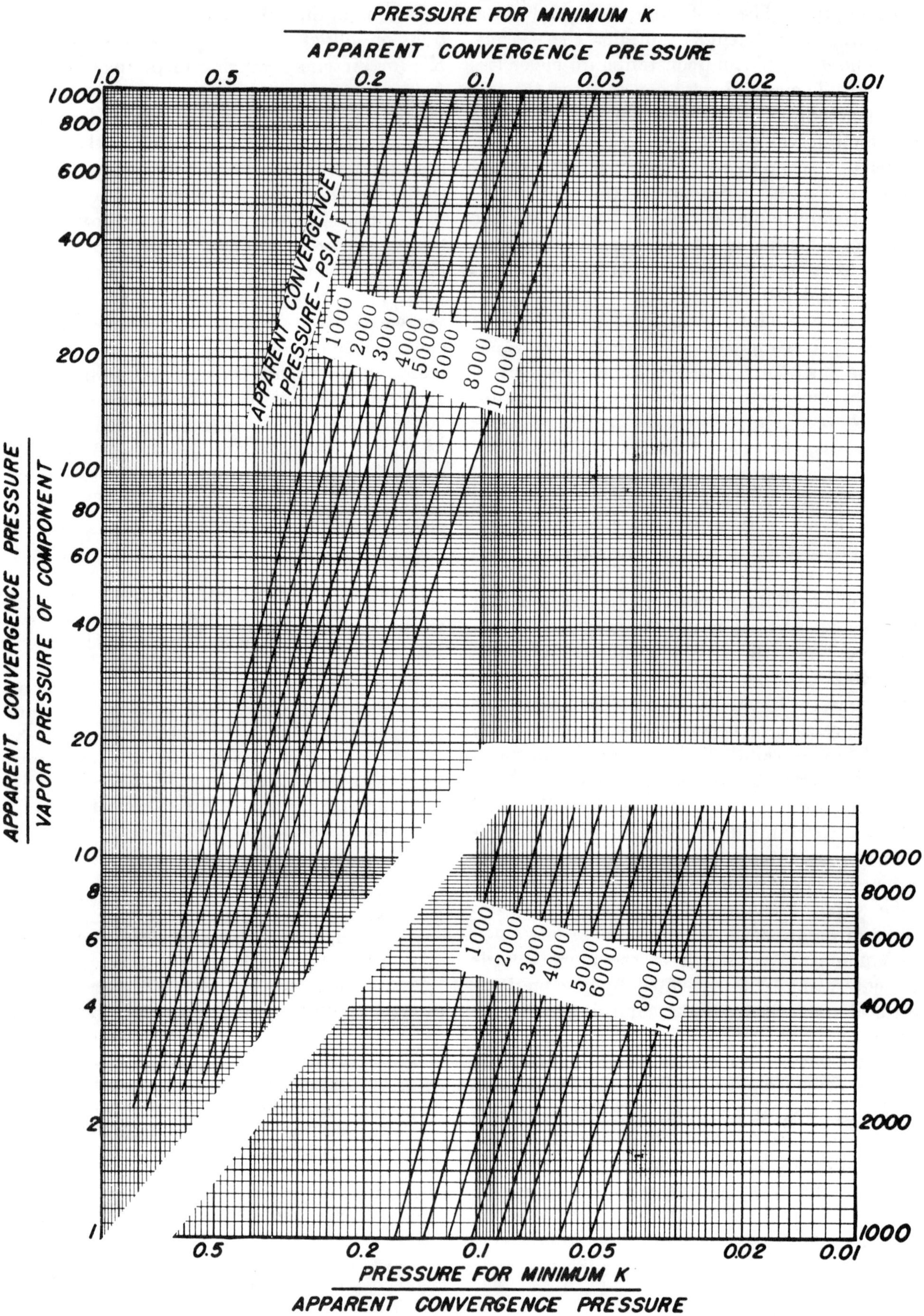

Figure 31—Correlation of the Pressure at Which the Minimum Equilibrium Constant Occurs with the Ratio of Apparent Convergence Pressure to the Vapor Pressure of the Component

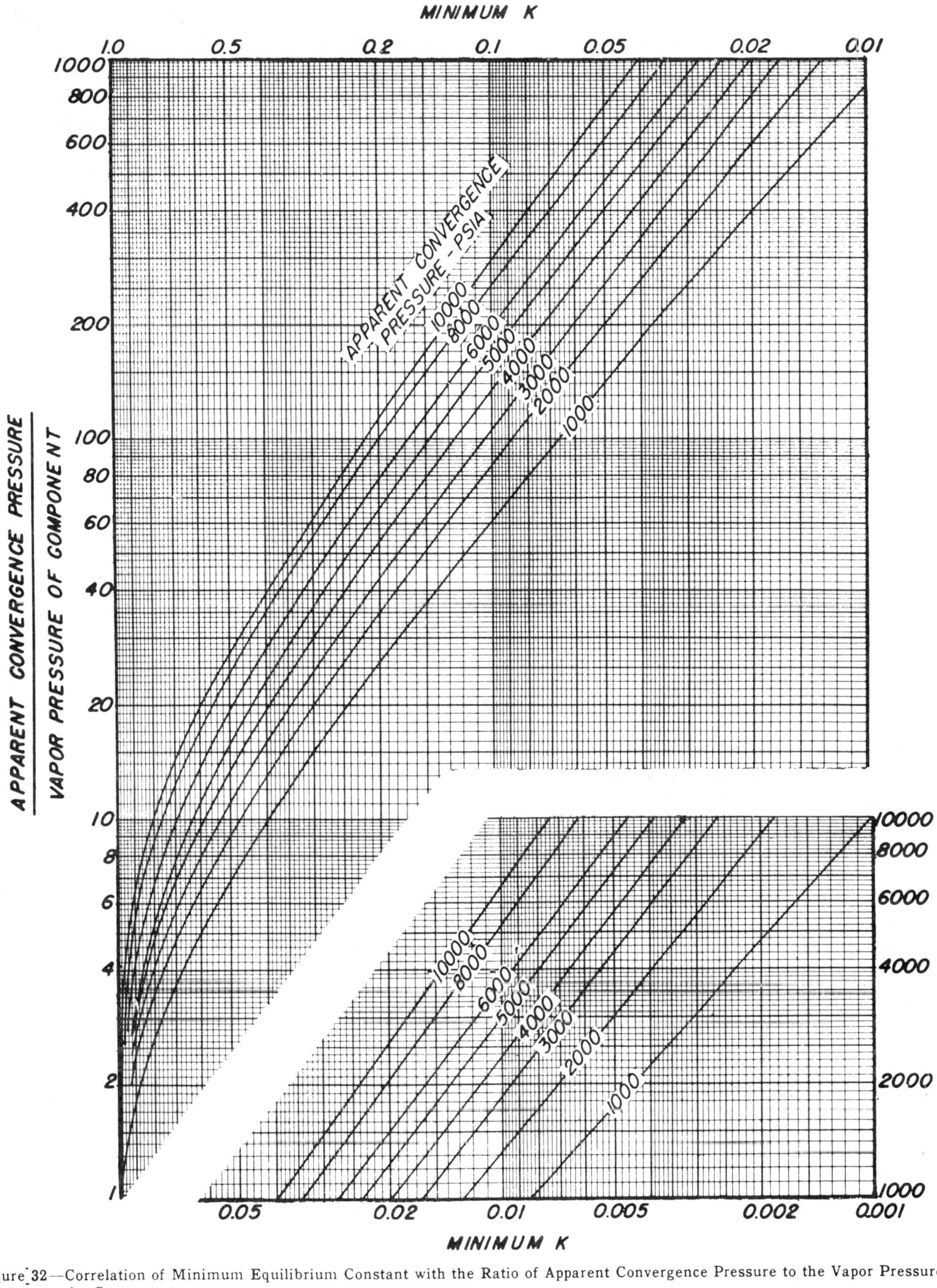

Figure 32—Correlation of Minimum Equilibrium Constant with the Ratio of Apparent Convergence Pressure to the Vapor Pressure of the Component

plotted at the experimentally determined bubble point pressure. Curves are sketched in through these values to give an approximate apparent convergence pressure.

(c) The locations of the minimum K values are determined by use of Figures 31 and 32, the estimated apparent convergence pressure, and vapor pressure values of the individual components.

(d) Smooth curves are drawn that connect the apparent convergence pressure, the bubble point pressure, the minimum K values, and the vapor-pressure points. In the low pressure region the curves are given slopes of 45 degrees.

(e) If the curves do not look right, the apparent convergence is changed slightly and another set of curves is constructed. This is continued until a set is obtained that has the appearance of those illustrated in Figure 30.

(f) The methane equilibrium constant curve is constructed parallel to the unmodified curve at pressures below 1000 psia. The curve is brought into the apparent convergence pressure, maintaining the form illustrated in Figure 30.

(2) When PVT data are not available to give the bubble point pressure, apparent convergence pressures are assumed roughly as follows:

Molecular Weight of C_7^+ Fraction	Apparent Convergence Pressure (psia)
120	3000
170	6000
220	9000

Admittedly, the above method has many faults. However, in the region of pressures wherein the system composition influences the equilibrium constants, it is better to use such a method of taking into account the system composition than to neglect the effect completely.

Calculation of Vapor-Liquid Equilibrium

The use of equilibrium constants makes possible the calculation of bubble point pressures, dew point pressures, and the proportions of vapor and liquid in equilibrium at pressures and temperatures where two phases can exist. The mathematical equations used in making these calculations are derived in the following section.

Flash Vaporization of a System. In all calculations it is assumed that the system is in thermodynamic equilibrium at the given temperature and pressure. The equations used for the calculation of the relative proportions of vapor and liquid are obtained from a material balance, as follows:

If we let

F = total moles in the system
V = total moles of gas at equilibrium
L = total moles of liquid at equilibrium
x = mole fraction of any component in the liquid
y = mole fraction of any component in the gas
n = mole fraction of any component in the system
$K = \frac{y}{x}$ = equilibrium constant for any component

A material balance of the system gives:

$$F = L + V \tag{23}$$

A material balance of any component, k, gives:

$$n_k F = x_k L + y_k V \tag{24}$$

Substituting $x_k K_k$ for y_k in Equation (24) and solving for x_k gives the mole fraction of the component in the liquid phase:

$$x_k = \frac{F n_h}{L + V K_k} \tag{25}$$

At equilibrium, the mole fractions of the components in both phases must sum to unity, or:

$$\sum_{k=1}^{k=j} x_k = 1 \tag{26}$$

$$\sum_{k=1}^{k=j} y_k = 1 \tag{27}$$

Applying the requirements of Equation (26) to Equation (25),

$$\sum_{k=1}^{k=j} x_k = \sum_{k=1}^{k=j} \frac{F n_k}{L + V K_k} = 1 \tag{28}$$

A similar equation can be obtained by solving for the composition of the vapor phase. The equation is

$$\sum_{k=1}^{k=j} y_k = \sum_{k=1}^{k=j} \frac{F n_k}{\frac{L}{K_k} + V} = 1 \tag{29}$$

Equations (28) and (29) must be solved by trial and error methods. The calculations are simplified by letting F equal 1, in which case the sum of L and V must equal 1. The procedure is to assume a value of L or V and calculate values of $\frac{n_k}{L + VK_k}$. By plotting the summation of $\frac{n_k}{L + VK_k}$ against the assumed L or V and using the resulting curve to guide the next assumption, a satisfactory solution can be obtained, usually by the third trial. This procedure is illustrated in Table 11 by the example of calculation of the quantities and composition of vapor and liquid in equilibrium at 2000 psia and 218°F.[37]

TABLE 11. EXAMPLE CALCULATION OF EQUILIBRIUM QUANTITIES OF GAS AND LIQUID
Pressure = 2000 *psia*, *Temperature* = 218°F

1	2	3	4	5	6	7
Component	Mole Fraction n	Equilibrum Constant at 2000 PSIA & 218°F K	Assume L=0.790 VK	L+VK	V=0.210 n/(L+VK)	Composition of Vapor y=Kx
C_1	0.4404	2.85	0.598	1.388	0.3173	
C_2	0.0432	1.17	0.246	1.036	0.0413	
C_3	0.0405	0.680	0.143	0.933	0.0434	
C_4	0.0284	0.410	0.0860	0.876	0.0324	
C_5	0.0174	0.244	0.0512	0.8412	0.0207	
C_6	0.0290	0.150	0.0315	0.8215	0.0353	
C_7+	0.4011	0.0160	0.00336	0.7934	0.5055	
	1.0000				0.9959	
			Assume L=0.776		V=0.224	
C_1			0.638	1.414	0.3114	
C_2			0.262	1.038	0.0416	
C_3			0.153	0.929	0.0436	
C_4			0.0918	0.8678	0.0327	
C_5			0.0546	0.8306	0.0210	
C_6			0.0336	0.8096	0.0358	
C_7+			0.00359	0.7796	0.5145	
					1.0006	
			Assume L=0.778		V=0.222	
C_1			0.632	1.410	0.3123	0.8896
C_2			0.260	1.038	0.0416	0.0487
C_3			0.151	0.929	0.0436	0.0296
C_4			0.0910	0.869	0.0327	0.0134
C_5			0.0542	0.8322	0.0209	0.0051
C_6			0.0333	0.8118	0.0357	0.0054
C_7+			0.00355	0.7815	0.5132	0.0082
					1.0000	1.0000

The calculations outlined in Table 12 and summarized in Figure 33 illustrate a typical problem concerned with the separation of gas and liquid by surface traps. In Table 12 each stage separation involves the application of Equation (28) to the material entering that stage. From experience it has been found easiest to treat each stage separately on the basis of one mole input to that stage. The actual quantities of vapor and liquid are calculated later from the results of the individual computations.

Bubble Point Pressure of a System. The bubble point of a system has been defined as the state at which an infinitesimal quantity of gas is in equilibrium with a large quantity of fluid. This definition leads to an equation for calculating the bubble point pressure of a system. Starting with Equation (29), it can be seen that as the vapor quantity becomes infinitesimal,

$$V \to O, \quad L \to F, \text{ so that}$$

$$\sum_{k=1}^{k=j} y_k = \lim_{V \to O} \sum_{k=1}^{k=j} \frac{Fn_k}{\frac{L}{K_k} + V} = \sum_{k=1}^{k=j} K_k n_k = 1 \tag{30}$$

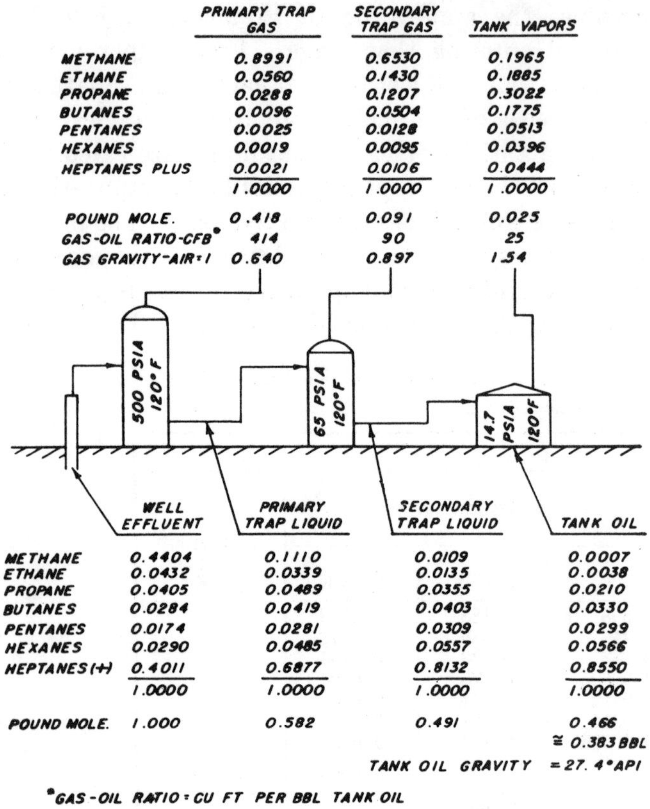

Figure 33—Stage Separation of Well Production

Thus, to calculate the bubble point pressure of a system, it is necessary to determine by trial and error methods the pressure at which the summation of the products of the mole fractions of the entire system times the equilibrium constants is equal to unity. An example calculation illustrating this procedure is shown in Table 13. The composition of the bubble point vapor is given by the values of $K_k n_k$ in column 4.

Dew Point Pressure of a System. The equation for the dew point condition can be developed from Equation (28). Because the liquid phase is infinitesimal at the dew point state,

$$L \to O, \qquad V \to F, \text{ so that}$$

$$\sum_{k=1}^{k=j} x_k = \lim_{L \to O} \sum_{k=1}^{k=j} \frac{F n_k}{L + V K_k} = \sum_{k=1}^{k=j} \frac{n_k}{K_k} = 1 \qquad (31)$$

The evaluation of the dew point pressure requires a trial and error solution in the same manner as the bubble point pressure. The values of n_k/K_k give the composition of the dew point liquid.

Accuracy of Vapor-Liquid Calculations

The accuracy of calculated vapor and liquid quantities, using equilibrium constants and the over-all system composition, is dependent on the composition of the system, the relative proportions of vapor and liquid in equilibrium, and errors in the individual equilibrium constants. Because of the several variables, it is impossible to make a general statement regarding accuracy, but examples of several individual cases will illustrate a method of evaluating the effects of the variables.

The calculations in Table 14 show the effect that an error in any one equilibrium constant can have on the quantity of liquid phase computed by use of Equation (28). The dissolved gas system having the composition shown in column 2 has been shown in Table 11 to form 0.778 mole fraction liquid and 0.222 mole fraction vapor at 2000 psia and 218°F. The equilibrium constants used to calculate these quantities are shown in column 3. Columns 4, 5, and 6 pertain to an erroneous calculation that assumes 0.770 mole fraction of the system is liquid. The fact that the summation of the mole fraction of the components in the liquid phase does not equal unity proves that $L = 0.770$ is an incorrect assumption. The adjusted mole fractions, x', shown in column 7, are the values to which each component would have to be changed in order that $L = 0.770$ be judged a correct value. For example, if the mole fraction of methane in the liquid phase were 0.3064, rather than 0.3089, the summation of the mole fraction in the liquid phase would have been unity, and $L = 0.770$ would have been a valid assumption. Likewise, the conditions

$$\sum_{k=1}^{k=j} x_k = 1$$

would be satisfied if the ethane liquid fraction were 0.0025 unit lower. Columns 8 and 9 present the calculation of $L + VK'$ and VK', resulting from values of x' shown in column 7. The equilibrium constant, K' in column 10 is the value which, used with the correct constants of the other components, would cause a calculated liquid quantity of 0.770 mole fraction, rather than the correct value of 0.778 mole fraction. Thus, for the particular system at 2000

TABLE 12. EXAMPLE OF TRAP SEPARATION OF A WELL EFFLUENT

Primary Trap, 500 PSIA, 120° F

1	2	3	4	5
Component	Well Effluent Mole Fraction n	Equilibrium Constant at 500 PSIA & 120°F K	Assume L=0.582 V=0.418 n/(L+VK)	y=Kx
C_1	0.4404	8.1	0.1110	0.8991
C_2	0.0432	1.65	0.0339	0.0560
C_3	0.0405	0.59	0.0489	0.0288
C_4	0.0284	0.23	0.0419	0.0096
C_5	0.0174	0.088	0.0281	0.0025
C_6	0.0290	0.039	0.0485	0.0019
C_7+	0.4011	0.0030*	0.6877	0.0021
	1.0000		1.0000	1.0000

Tank, 14.7 PSIA, 120° F

1	2	3	4	5
Component	Secondary Trap Liquid Mole Fraction n	Equilibrium Constant at 14.7 PSIA & 120°F K	Assume L=0.9486, V=0.0514 n/(L+VK)	y=Kx
C_1	0.0109	265	0.0007	0.1965
C_2	0.0135	46.5	0.0038	0.1885
C_3	0.0355	14.3	0.0210	0.3022
C_4	0.0403	5.35	0.0330	0.1775
C_5	0.0309	1.72	0.0299	0.0513
C_6	0.0557	0.70	0.0566	0.0396
C_7+	0.8132	0.052*	0.8550	0.0444
	1.0000		1.0000	1.0000

Moles tank oil per mole well effluent
= 0.491 · 0.949 = 0.466

Moles tank vapors per mole well effluent
= 0.491 · 0.051 = 0.025

Secondary Trap, 65 PSIA, 120° F

1	2	3	4	5
Component	Primary Trap Liquid Mole Fraction n	Equilibrium Constant at 65 PSIA & 120°F K	Assume L=0.844 V=0.156 n/(L+VK)	y=Kx
C_1	0.1110	60.0	0.0109	0.6530
C_2	0.0339	10.6	0.0135	0.1430
C_3	0.0489	3.4	0.0355	0.1207
C_4	0.0419	1.25	0.0403	0.0504
C_5	0.0281	0.415	0.0309	0.0128
C_6	0.0485	0.170	0.0557	0.0095
C_7+	0.6877	0.013*	0.8132	0.0106
	1.0000		1.0000	1.0000

Moles secondary trap liquid per mole well effluent
= 0.582 · 0.844 = 0.491

Moles secondary trap gas per mole well effluent
= 0.582 · 0.156 = 0.091

1	2	3	4
Component	Tank Oil Mole Fraction n	Pound Mole Per Gallon at 14.7 PSIA & 60°F	Gallons per Mole Tank Oil
C_1	0.0007	0.129	——
C_2	0.0038	0.104	0.04
C_3	0.0210	0.0958	0.22
C_4	0.0330	0.0821	0.40
C_5	0.0299	0.0724	0.41
C_6	0.0566	0.0641	0.88
C_7+	0.8550	0.0263	32.53
	1.0000		34.48

Barrels tank oil per pound mole well effluent =

$$\frac{0.466 \cdot 34.48}{42} = 0.383$$

Molecular Weight C_7+ = 287
Specific Gravity C_7+ = 0.9071

* Equilibrium constant of Heptanes(+) assumed to be 0.15 times that of Heptane.[33]

psia and 218°F, a +1.8 per cent error in the methane constant would cause a −1.0 per cent error in the calculated liquid quantity. To cause the same −1.0 per cent error in the calculated liquid requires that the pentane equilibrium constant be in error by +195.1 per cent.

Results of calculations similar to those presented in Table 14 are shown in Figures 34, 35, and 36. The curves of Figure 36 are of interest because they show that calculations near the dew point pressure of a condensate system can be influenced greatly by both the methane and heptanes plus equilibrium constant. For the particular conditions under study in Figure 36, an error of −2.5 per cent in the value of the methane equilibrium constant results in an error of +40 per cent in the quantity of calculated liquid. An error of +8 per cent in the same constant results in the calculation of a dew point state, rather than the correct 0.051 mole fraction liquid content. Errors in the heptanes plus constant, while not as drastic in their effect on calculated liquid volumes as errors in the methane constant, also cause large errors in the calculated phase quantities.

Table 14. Effect of Errors in Equilibrium Constants on the Calculated Quantity of Liquid Phase in a Dissolved Gas System

1	2	3	4	5	6	7	8	9	10	11
			Assume L = 0.770 V = 0.230							
Component	System Composition Mole Fraction n	Equilibrium Constant 2000 PSIA & 218°F K	VK	L+VK	Liquid Phase Composition x	Adjusted Liquid Phase Composition x′	L+VK′	VK′	K′	Error in Equilibrium Constant Per Cent
C_1	0.4404	2.850	0.6555	1.4255	0.3089	0.3064	1.4373	0.6673	2.901	+1.8
C_2	0.0432	1.170	0.2691	1.0391	0.0416	0.0391	1.1048	0.3348	1.456	+24.4
C_3	0.0405	0.680	0.1564	0.9264	0.0437	0.0412	0.9830	0.2130	0.926	+36.2
C_4	0.0284	0.410	0.0943	0.8643	0.0328	0.0303	0.9372	0.1672	0.727	+77.3
C_5	0.0174	0.244	0.0561	0.8261	0.0211	0.0186	0.9355	0.1655	0.720	+195.1
C_6	0.0290	0.150	0.0345	0.8045	0.0360	0.0355	0.8656	0.0956	0.416	+177.3
C_7+	0.4011	0.016	0.0037	0.7737	0.5184	0.5159	0.7775	0.0075	0.033	+106.2
	1.0000				1.0025					

NOTE: Correct liquid quantity is 0.778 mole fraction.

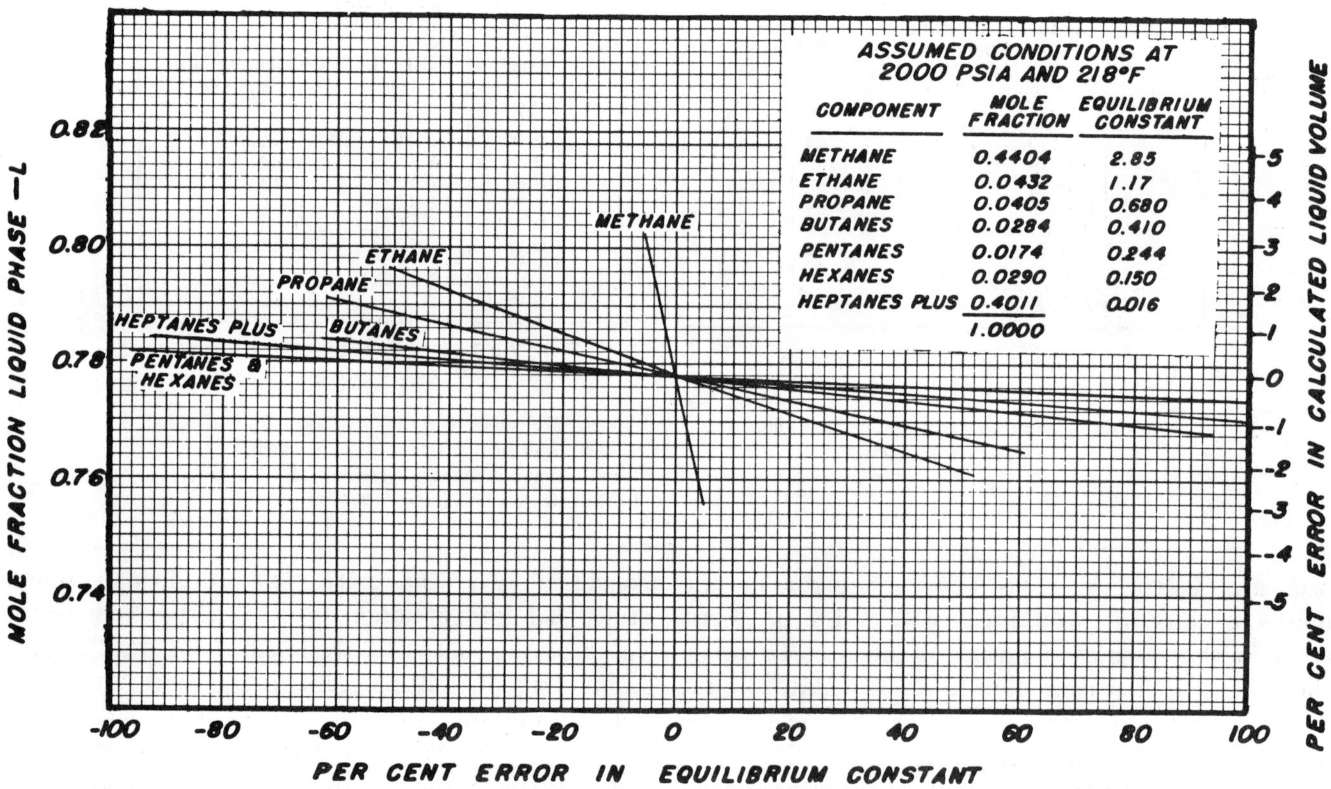

Figure 34—Effect of Errors in Equilibrium Constants on the Calculated Quantity of Liquid Phase in a Dissolved Gas System

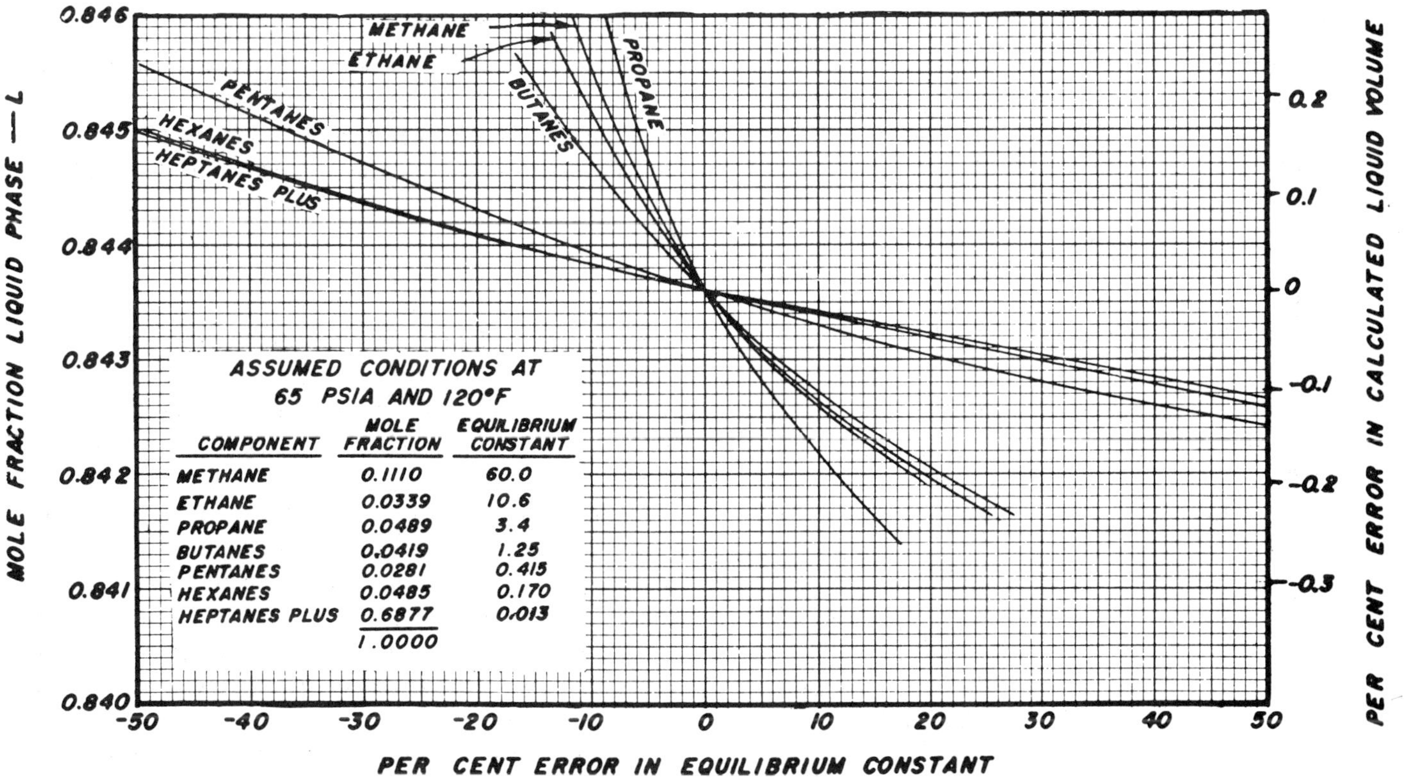

Figure 35—Effect of Errors in Equilibrium Constants on the Calculated Quantity of Liquid Phase in a Dissolved Gas System

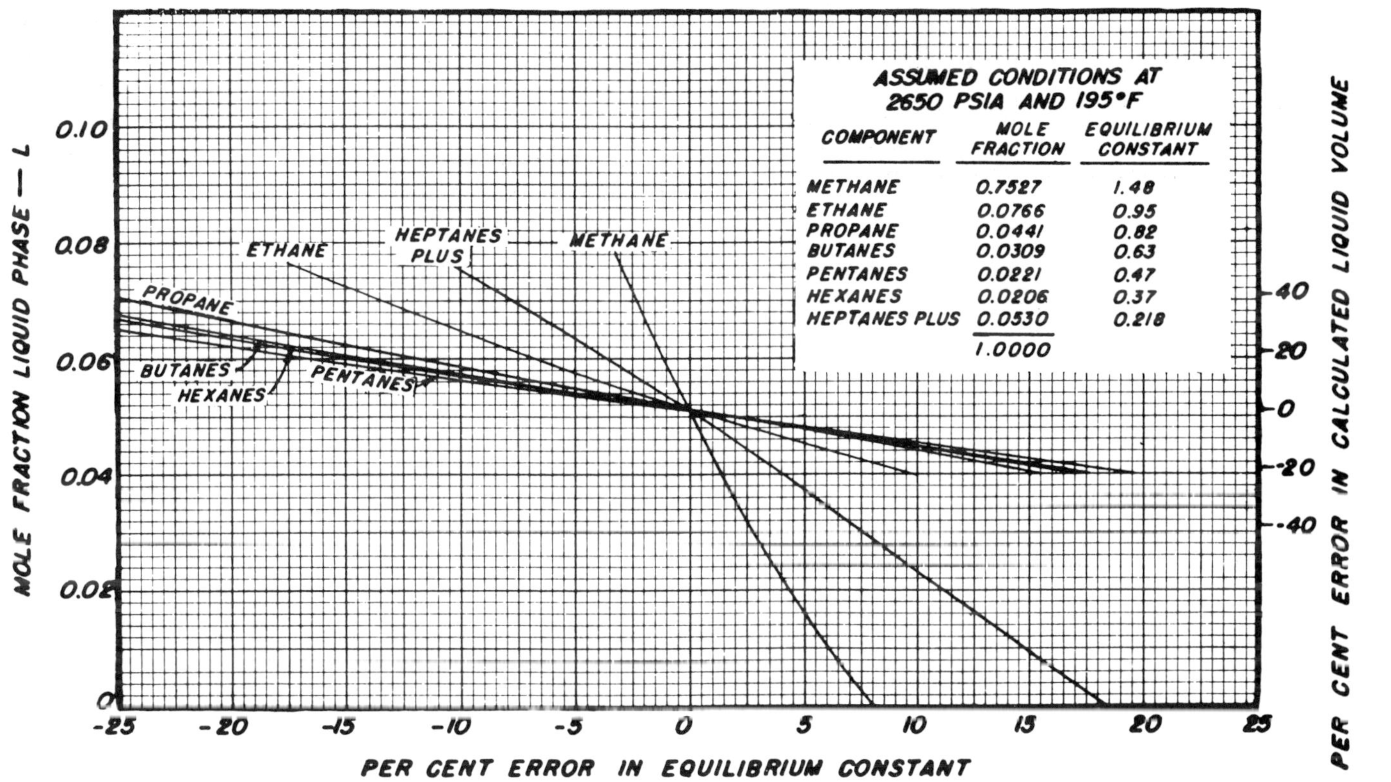

Figure 36—Effect of Errors in Equilibrium Constants on the Calculated Quantity of Liquid Phase in a Gas-Condensate System

TABLE 13. EXAMPLE CALCULATION OF THE BUBBLE POINT PRESSURE OF A WELL EFFLUENT AT 218°F

1	2	3	4
	Well Effluent Mole Fraction	Assume Pressure P = 3000 PSIA	
Component	n	Equilibrium Constant K	Kn
C_1	0.4404	2.15	0.9469
C_2	0.0432	1.03	0.0445
C_3	0.0405	0.672	0.0272
C_4	0.0284	0.440	0.0125
C_5	0.0174	0.300	0.0052
C_6	0.0290	0.215	0.0062
C_7+	0.4011	0.0235	0.0094
			1.0519
		Assume Pressure P = 3200 PSIA	
C_1		2.06	0.9072
C_2		1.025	0.0443
C_3		0.678	0.0274
C_4		0.448	0.0127
C_5		0.316	0.0055
C_6		0.230	0.0067
C_7+		0.0250	0.0100
			1.0138
		Assume Pressure P = 3300 PSIA	
C_1		2.02	0.8896
C_2		1.02	0.0441
C_3		0.680	0.0275
C_4		0.450	0.0128
C_5		0.323	0.0056
C_6		0.239	0.0069
C_7+		0.0260	0.0104
			0.9969

By interpolation, bubble point pressure = 3280 PSIA.

Summary of Principal Points

The equilibrium constants, K, used in calculating the relative proportions and compositions of vapor and liquid in equilibrium are functions of pressure, temperature, and over-all composition of the system. As yet, too few systems have been investigated to determine the manner in which the system composition influences the equilibrium constants.

At pressures less than 1000 psia, the constants determined by Katz and Hachmuth are best for calculating the behavior of dissolved gas systems. Above 1000 psia, they may require some adjusting in order to be consistent with experimentally determined phase behavior.

Condensate systems are best handled by use of Roland, Smith, and Kaveler's constants. Major revisions of the constants may be required to give correct phase behavior in the high pressure region.

The fact that log K-log P isotherms appear to converge to unity should not be interpreted as meaning that the system is at its critical state. Equilibrium constants can converge only at the critical temperature of the system; at all other temperatures the convergence is an "apparent convergence."

Calculations of bubble point pressures and the relative phase volume near bubble point conditions are most sensitive to inaccuracies of the methane equilibrium constant. Calculations at or near dew point conditions are sensitive to inaccuracies of both methane and heptanes plus constants.

CHAPTER 6

GAS-CONDENSATE SYSTEMS

For purposes of this discussion, a gas-condensate system is one which will exhibit isothermal retrograde behavior in the temperature range that is of interest to petroleum engineering studies. Such a system has a critical temperature lower than the reservoir temperature or the surface temperature at which it is separated into gas and liquid. Production of a condensate system through surface traps usually results in gas-oil ratios greater than 5,000 cu ft per bbl* and tank oil gravity between 45 and 70° API. The tank oils usually are white or only slightly colored. A definition of the upper value of gas-oil ratio of a gas-condensate system is arbitrary; usually a system having a ratio greater than 100,000 cu ft per bbl (0.43 gal per MCF) is considered a dry gas.

Pressure decline in gas-condensate reservoirs results in formation of a liquid phase within the reservoir rock pores. During the last fifteen years, the operation of processing production from gas-condensate reservoirs for the heavier hydrocarbon components and returning the light components to the reservoir (cycling) has become increasingly popular. It often is stated that the purpose of cycling is to maintain the reservoir pressure near its original pressure in order to remove the heavier hydrocarbon components which otherwise would form the retrograde liquid.

The PVT behavior of gas-condensate systems that are representative of reservoir fluids can be determined easily by laboratory tests. Information that usually is determined in the laboratory includes compressibility factors of the reservoir vapor phase and surface gases, quantity of liquid formed at reservoir and atmospheric conditions as a result of retrograde action, and data from which the decline in reservoir pressure resulting from production of reservoir vapor can be calculated directly. Less is known of the composition of reservoir vapor and retrograde liquid during pressure decline than of their volumetric behavior. Laboratory experiments to determine the composition of the two phases are laborious and time-consuming and have received little attention. Calculated compositions and phase volumes based solely on equilibrium constant data can be greatly in error. Probably the best estimate of phase composition can be obtained by combining laboratory-determined volumetric data with published equilibrium constant data. The phase composition so computed will not be greatly in error and will indicate the trend that the reservoir phases will follow.

* The figure 5,000 cu ft per bbl assumes production of only vapor from the reservoir. The production of both vapor and liquid from a reservoir to give 5,000 cu ft per bbl (or more) does not necessarily indicate a condensate system.

Laboratory Tests

The laboratory apparatus used in condensate studies differs from the apparatus used in dissolved gas studies for two reasons:

(1) The dew point pressure of most condensate systems cannot be detected by a sharp change in the pressure-volume relation of the system. (Bubble point pressure is determined most easily by this means.)

(2) The liquid phase constitutes only a small part of the total volume in the laboratory testing equipment. It is necessary therefore to have methods of precisely measuring small liquid quantities.

One of the most satisfactory PVT cells is one with glass windows to permit the visual observation of dew points and the quantity of retrograde liquid at pressures less than the dew point pressures. The type of cell in which the liquid phase volume is determined by locating the interface between vapor and liquid by thermal methods* is equally satisfactory when liquid volumes greater than two cubic centimeters are involved. However, the second type of cell is inferior to a glass cell when an accurate measure is desired of a system's dew point pressure, because the dew point pressure must be obtained by extrapolation of a liquid phase volume-pressure curve to zero liquid volume.

Flash Tests. Flash tests are carried out on gas-condensate systems to determine volumetric behavior of the systems under surface trap conditions, under well tubing conditions, and under reservoir conditions. As previously defined, a flash test signifies constant system composition during the progress of the test. The tests are made most often on systems prepared from surface samples of trap fluids recombined in their produced ratios.

It is not difficult to measure phase volumes in a laboratory PVT cell at the higher pressures corresponding to well tubing pressures and to reservoir pressures. The weight of hydrocarbon material in the cell during the tests can be determined accurately at the completion of the tests by measuring the quantity and gravity of the gas and liquid removed from the cell. Often the gas and liquid are removed from the cell under the pressure and temperature conditions existing in the field traps. When this is done, the volumetric data can be given in terms of formation volumes, or in terms of volumes under reservoir conditions per MCF of surface trap gases. For many purposes a specific volume or density value obtained from the measured weight of the system and its measured volume at specified pressures and temperatures is more valuable in engineering calculations than the formation volume or "reservoir" volume.

It is not always feasible to determine experimentally the effect of surface trap pressures and temperatures on the properties and volumes of surface products. Size limitations of the usual laboratory PVT equipment often make it impossible to expand high gas-oil ratio systems to pressures of the order of 250 psia and still have sufficient liquid phase in the cell to measure accurately its volume. Therefore, in many condensate systems, it is necessary to use equilibrium calculation methods to evaluate the effect of surface trapping conditions on the quantity and gravity of tank oil and trap gases. Because of the inaccuracies of the calculation methods, the absolute values obtained by this process may not agree with field measurements, although the trend shown by the calculation usually is correct.

Differential Tests. By definition, a differential process is one in which the over-all composition of the system varies as the process proceeds. The differential tests made on gas-condensate systems are based on the assumption that the retrograde liquid formed in the reservoir rock is immobile. In line with this assumption, the laboratory test procedure usually is as follows:

(1) A system having a composition as close as possible to that of the original reservoir fluid is constructed in the laboratory PVT cell. The system is brought to the original reservoir pressure and temperature or to the system's dew point pressure at the reservoir temperature.

(2) Starting at the dew point or reservoir pressure, equilibrium vapor phase material is removed from the top of the cell. At several points during the process, measurements are made of the cell pressure, the volume of retrograde liquid formed in the cell, and the cumulative volume of vapor removed from the cell. This step simulates the production of vapor phase from the reservoir rock.

(3) During the vapor removal process, the total volume occupied by the system is kept constant. This is equivalent to assuming that the reservoir volume does not change during its producing life because of water encroachment. The assumption, of course, may be invalid for many reservoirs, but its invalidity does not detract from the usefulness of the laboratory data because adjustments can be made to the laboratory data where the true reservoir condition is known.

The fundamental assumption of the differential tests concerning immobility of retrograde reservoir liquid is believed reasonable in view of the facts that

* This type cell and its operation are described in more detail in Chapter 7 and by Sage and Lacey.[38]

most condensate systems reported in the literature yield less than 10 per cent retrograde liquid by volume during pressure decline at reservoir temperatures, and that the effective permeability of reservoir rocks to this phase usually is zero in the saturation range of interest.

There is good reason to believe, however, that the above assumption is not valid in the region of the well bore. A higher liquid saturation probably exists near the well bore than out in the reservoir because of the larger pressure gradient and the increased fluid flux density near the well bore. Over a period of time the liquid saturation near the well bore may increase to the point where the pressure gradient may overcome the capillary forces retaining the liquid in the rock. The liquid phase will then be produced in the well concurrently with vapor material. This local action in the region of the well bore is of interest primarily to the problem of recombining surface samples to obtain a system that is representative of the system in the major portion of the reservoir.

Condensate System Behavior in the Single-Phase Region

The volumetric behavior of a condensate system under pressure and temperature conditions that permit only a single phase is similar to the behavior of dry gases discussed in Chapter 3. The volume of the system can be calculated by use of the gas law equation or by any one of several empirical correlations. The sections to follow will outline methods of evaluating compressibility factors for condensate system calculations, as well as present several empirical correlations.

Compressibility Factors. The volume occupied by a condensate system at pressures and temperatures outside the phase diagram can be calculated from the equation $PV = NZRT$. The compressibility factor, Z, can be determined from the pseudo-reduced pressure and temperature correlations shown in Figure 14, provided that proper pseudo constants are used to represent the relatively large proportion of high molecular weight components present in the system. Standing and Katz[13] have correlated pseudo-critical values and the molecular weight of the heptanes plus fraction of twelve high-pressure equilibrium vapors. The pseudo-critical pressures were found to be roughly 50 psi higher than true critical pressures of equivalent molecular weight normal paraffin hydrocarbon. The pseudo-critical temperatures were found to be about 30°F higher than the true critical temperatures of equivalent molecular weight paraffin hydrocarbons. Mathews, Roland, and Katz[14] found that volumetric calculations could be made with increased accuracy if the specific gravity, as well as the molecular weight, were used in evaluating the pseudo-critical constants of heptanes plus fractions. The correlation presented by these authors is reproduced in Figure 37 and is recommended for use where both the specific gravity and molecular weight of the heptanes plus fraction are known.

Where neither specific gravity nor molecular weight is known, it is necessary to assume pseudo-critical values. A reasonable assumption is that the molecular weight of the heptanes plus is equivalent to that of nonane, which is 128. By adding the 50 psi and 30°F found by Standing and Katz to nonane's true criticals, the pseudo-critical values are 385 psia and 1100°F.

The compressibility factor, Z, as normally used in the equations $PV = NZRT$ or $\overline{V} = \frac{ZRT}{PM}$ should be applied only in the pressure and temperature range in which the system remains in a single phase. If the equation is used at pressures and temperatures where two phases exist, $\overline{V}$ in the second equation expresses the average specific volume of the two-phase system. In many instances, the average specific volume of the system when in two phases is close to the specific volume calculated as though the system existed as a single phase. This is particularly true in the pressure interval 2000 to 6000 psia when the liquid phase represents less than 5 per cent of the total system volume.

Pseudo-critical Constant-Gravity Relations of Condensate Systems. The basis for correlations of the pseudo-critical pressure and temperature with molecular weight of a gas has been explained in Chapter 3. The correlations found by Standing and Katz[13] for a group of condensate vapors are presented in Figure 17 and should be used in calculations pertaining to the volumetric properties of condensate systems.

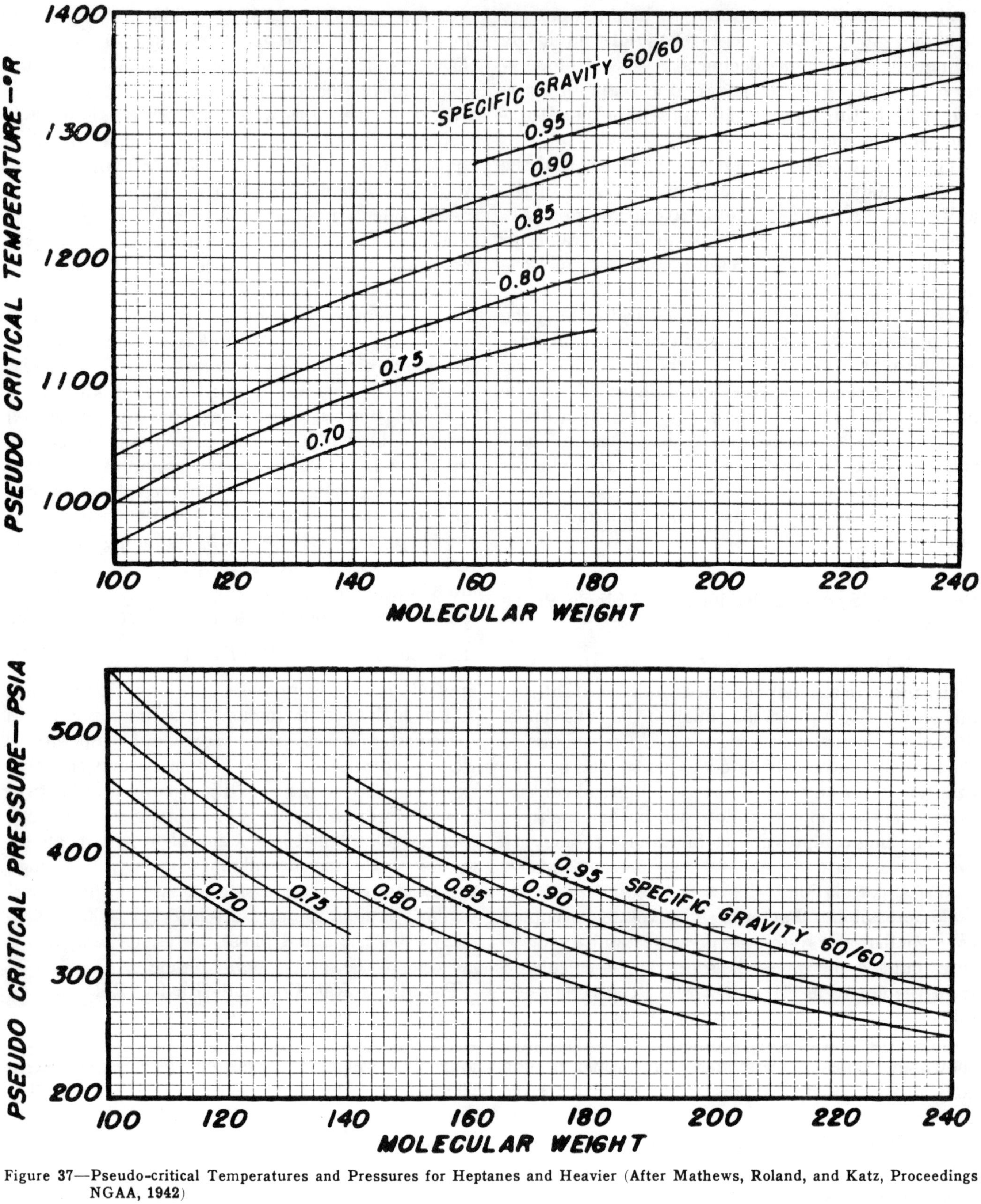

Figure 37—Pseudo-critical Temperatures and Pressures for Heptanes and Heavier (After Mathews, Roland, and Katz, Proceedings NGAA, 1942)

In an effort to simplify condensate system calculations, Rzasa and Katz[39] developed a chart that shows a relationship between the ratio of well fluid gravity (as a vapor) to trap gas gravity and the barrels of surface condensate per million cubic feet of surface gas. Such a chart has considerable utility, as one often knows the surface trap gas gravities, but for calculation purposes requires the gravity of the system in the well or in the reservoir rock. Rzasa and Katz' correlation contained only a single curve, but theoretically such a correlation also should contain variables of trap gas gravity, condensate liquid gravity, and condensate molecular weight. These variables are related by the equation:

$$\frac{\text{well fluid gravity}}{\text{trap gas gravity}} = \frac{M_1}{28.97\gamma_g}\left(\frac{76.4S\gamma_g + 350\delta_1}{2.64SM_1 + 350\delta_1}\right) \tag{32}$$

where
M_1 = molecular weight of condensate (tank oil)
δ_1 = specific gravity of condensate (tank oil)
γ_g = gravity of trap gas (air = 1)
S = MCF surface gas per barrel condensate

Fortunately, there exists a fair correlation of molecular weight with the specific gravity of natural condensates so that Equation (32) can be simplified to contain only data which can be determined in the field. The result of such simplifications is the calculating chart presented in Figure 38. The rough correlation of condensate molecular weight and gravity used also is shown in Figure 38. An example problem using the data presented in Figure 38 is given in Table 15.

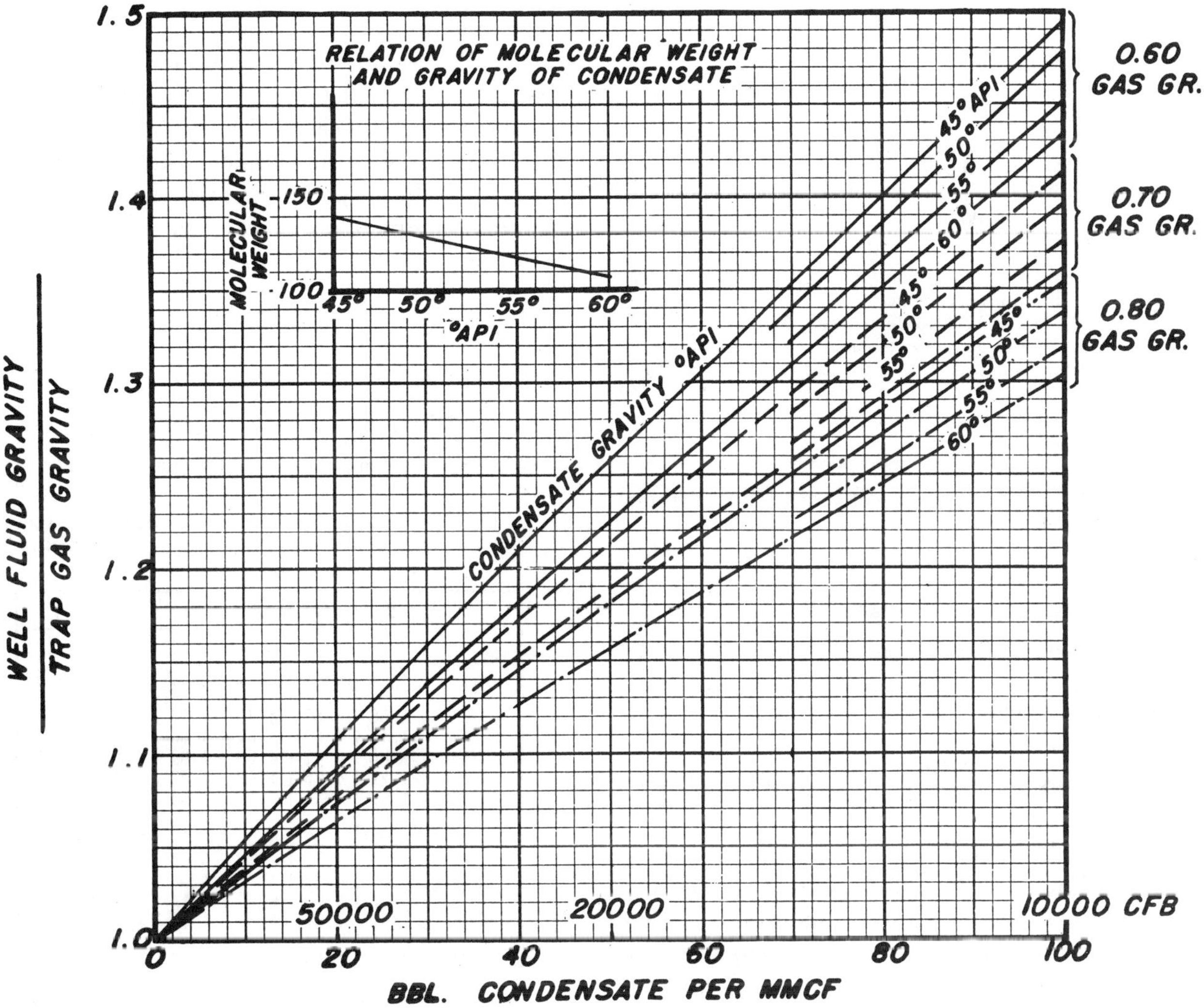

Figure 38—Effect of Condensate Volume on the Ratio of Surface Gas Gravity to Well Fluid Gravity

TABLE 15. ESTIMATION OF GAS-IN-PLACE AT ELEVATED PRESSURE AND TEMPERATURE USING WELL GRAVITY TO EVALUATE PSEUDO-CRITICAL CONSTANTS

Problem: To estimate the standard cubic feet of gas in place in a reservoir that has the following properties:

Given:		
	Reservoir pressure	3000 psia
	Reservoir temperature	240°F
	Rock porosity	30 per cent
	Interstitial water	27 per cent
	Tank oil production	323 bbl per day
	Tank oil gravity	45° API
	Primary trap gas rate	3,765 MCF per day
	Primary trap gas gravity	0.65 (air = 1)
	Tank vapor rate	169.2 MCF per day
	Tank vapor gravity	1.25 (air = 1)

Solution: Basis: One acre-foot sand volume

Average separator gas gravity

$$= \frac{3{,}765 \cdot 0.65 + 169 \cdot 1.25}{3{,}765 + 169} = 0.677$$

$$\text{Bbl condensate per MMCF} = \frac{323}{3.934} = 82.1$$

$$\text{From Figure 38, } \frac{\text{well fluid gravity}}{\text{trap gas gravity}} = 1.35$$

Well fluid gravity = 1.35 · 0.677 = 0.917

From Figure 17, $pT_C = 430°$ R, $pP_C = 650$ psia

$$pT_R = \frac{460 + 240}{430} = \frac{700}{430} = 1.63 \qquad pP_R = \frac{3000}{650} = 4.62$$

From Figure 15, $Z = 0.840$

The reservoir hydrocarbon space per acre-foot of rock is 43560 0.30 · (1 − 0.27) = 9,540 cu ft

Total moles of hydrocarbons per acre-foot at reservoir conditions,

$$N = \frac{PV}{ZRT} = \frac{3000 \cdot 9540}{0.840 \cdot 10.73 \cdot 700} = 4540 \text{ lb moles}$$

Now, if all this quantity of hydrocarbons were produced as gas at the surface, it would represent

4540 · 379 = 1730 MCF per acre-foot

However, as a portion will be produced as tank oil, it is necessary to make allowance for this fact. One barrel of 45° API tank oil (0.802 specific gravity) having a molecular weight of 140 would correspond to

$$\frac{5.615 \cdot 62.37 \cdot 0.802}{140} = 2.01 \text{ lb moles}$$

At the present producing ratio of 12,170 cu ft per bbl, the total surface gas amounts to

$$\frac{12{,}170}{379} = 32.1 \text{ moles}$$

Therefore, under the present producing ratio,

$$\frac{32.1}{32.1 + 2.01} = 0.941 \text{ mole fraction of the reservoir material}$$

will be surface gases, and 0.059 mole fraction will be tank oil.

Surface gas = 0.941 · 4540 · 379 = 1620 MCF per acre-foot

$$\text{Surface tank oil} = \frac{0.059 \cdot 4540}{201} = 133 \text{ bbl per acre-foot}$$

A. L. Vitter, Jr.[40] has suggested an equation similar to Equation (32) by which to correct the observed trap gas gravity to a well fluid gravity. By assuming that all condensates (tank oils) have an equivalent vapor volume of 200 cu ft of gas per cu ft of condensate, he obtained the following equation:

$$\gamma_w = \frac{\gamma_g + \dfrac{4591\delta_1}{1000S}}{1 + \dfrac{1123}{1000S}} \tag{33}$$

where

γ_w = gravity of well fluid (air = 1)

Well fluid gravities calculated by this equation do not differ materially from those calculated by use of Equation (32) or Figure 38. In the example given in Table 15, a well fluid gravity of 0.896 was obtained by Vitter's equation compared to 0.917 obtained by use of Figure 38. Better comparisons are obtained at high gas-oil ratios and high °API condensate gravities.

Empirical Behavior in the Single-Phase Region. Sage and Olds[8] correlated the volumetric behavior of five condensate systems produced from California fields. The producing gas-oil ratios of the systems were in excess of 5,000 cu ft per bbl, and the tank oil gravities were between 52° and 63° API. An empirical relation was determined which expressed the formation volume of the system as a function of pressure, temperature, and gas-oil ratio. The relation is

$$u = A\frac{rT}{P} \tag{34}$$

where

A = empirical coefficient, as tabulated in Table 16
P = pressure (psia)
T = temperature (°R)
r = gas oil ratio, cu ft per bbl of tank oil
u = formation volume (bbl per bbl of tank oil)

TABLE 16. VALUES OF COEFFICIENT A USED TO CALCULATE FORMATION VOLUME OF GAS-CONDENSATE SYSTEMS, METHOD OF SAGE AND OLDS[8]*

1	2	3	4	5	6	7
Pressure PSIA	A · (10³)					
	100° F.	130° F.	160° F.	190° F.	220° F.	250° F.
600	4.58	4.67	4.75	4.83	4.89	4.93
800	4.46	4.57	4.67	4.76	4.83	4.87
1000	4.35	4.47	4.59	4.69	4.77	4.82
1250	4.21	4.35	4.49	4.62	4.71	4.77
1500	4.09	4.25	4.41	4.55	4.66	4.73
1750	3.99	4.17	4.34	4.50	4.62	4.71
2000	3.93	4.11	4.29	4.46	4.59	4.69
2250	3.89	4.08	4.26	4.43	4.57	4.68
2500	3.88	4.06	4.25	4.42	4.57	4.68
2750	3.89	4.07	4.26	4.44	4.58	4.69
3000	3.92	4.10	4.29	4.47	4.61	4.71

* For use in Equation (34).

Theoretically, A is not a function of pressure and temperature only, as indicated by Table 16. The gas-oil ratio, the gravity of the gas, oil gravity, and molecular weight are factors which cause minor variation in the value of A. Notwithstanding the apparent neglect of the above factors, the method gives formation value factors that follow the volumetric behavior of the California systems with errors no larger than those obtained when other methods are used. An example of use of this method is shown in Table 17.

An empirical correlation of the gas plus liquid phases formation volume has been made by Standing.[7] This correlation is satisfactory for dissolved

Table 17. Estimation of Gas-in-place at Elevated Pressure and Temperature Using Method of Sage and Olds[8]

Problem: *To estimate the standard cubic feet of gas in a reservoir having the following properties:*

Given:	Reservoir pressure	3000 psia
	Reservoir temperature	240°F
	Porosity	30 per cent
	Interstitial water	27 per cent
	Tank oil production	323 bbl per day
	Primary trap gas rate	3,765 MCF per day
	Tank vapor rate	169.2 MCF per day

Solution: Basis: One acre-foot sand volume

Producing gas-oil ratio

$$= \frac{(3765 + 169)(10^3)}{323} = 12{,}170 \text{ cu ft per bbl}$$

At 3000 psia and 240°F, $A = 4.68(10^{-3})$ (Table 16) Formation volume,

$$u = \frac{ArT}{P} = \frac{4.68(10^{-3}) \cdot 12{,}170 \cdot (460 + 240)}{3000}$$

$= 13.30$ bbl per bbl tank oil

Reservoir hydrocarbon volume occupied by one day's production is $13.30 \cdot 323 \cdot 5.615 = 24{,}100$ cu ft

One acre-foot reservoir rock volume has $43{,}560 \cdot 0.30 \cdot (1 - 0.27) = 9{,}540$ cu ft hydrocarbon space.

Therefore, one acre-foot sand contains $\frac{9{,}540}{24{,}100} = 0.396$ of a day's production. This is equivalent to $0.396(3765 + 169) = 1557$ MCF surface gas, and $0.396 \cdot 323 = 128$ bbl of tank oil

gas systems, as well as condensate systems. From the correlation functions shown in Figure 39, it can be seen that at high gas-oil ratios the effect of the tank oil gravity becomes negligible and the compressibility factor of the system then becomes a function of the trap gas gravity. The formation volumes calculated by use of this relation usually are in error less than 3 per cent. The advantage of the correlation is that it can be used easily in the form of a calculating chart.* The formation volume calculation in the example shown in Table 18 normally is carried out by use of the calculation chart.

* See Appendix I.

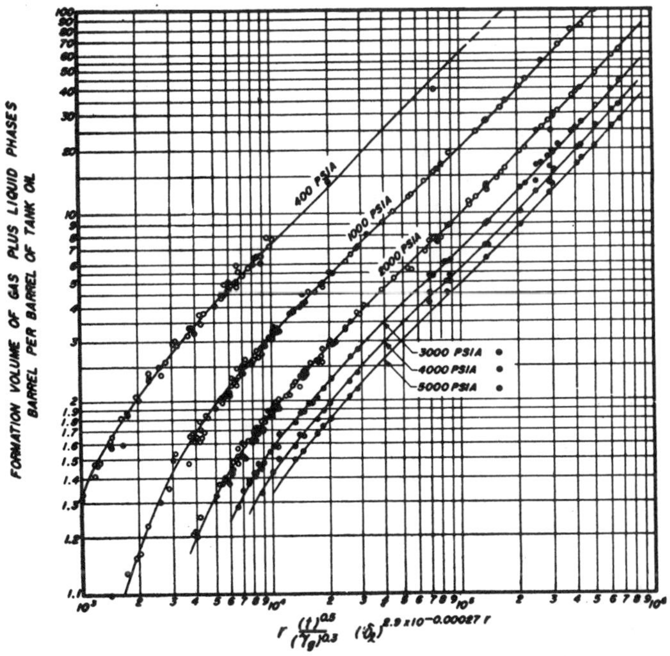

Figure 39—Formation Volume of Gas Plus Liquid Phases from Gas-Oil Ratio, Gas Gravity, Tank Oil Gravity, Temperature, and Pressure

Condensate System Behavior in the Two-Phase Region

Dew Point Behavior. The dew point, like the bubble point, defines the state at which a system changes from one phase to two phases. The dew point pressure can be calculated, provided the proper equilibrium constants and the composition of the system are known. While it is not too difficult to determine the composition of a system, the blind use of published equilibrium constants can result in errors as great as several thousand pounds per square inch in the calculated dew point pressures.

Sage and Olds[8] have presented a relation between tank oil gravity, gas-oil ratio, temperature, and dew point pressure for the five San Joaquin Valley, California, condensate systems previously mentioned. Their data, presented in Table 19, represent the only published attempt of a general correlation of dew point pressures. Although the five systems presented a satisfactory correlation within themselves,

TABLE 18. ESTIMATION OF GAS-IN-PLACE AT ELEVATED PRESSURE AND TEMPERATURE USING STANDING'S CORRELATION[7]

Problem: *To estimate the standard cubic feet of gas in place in a reservoir having the following properties:*

Given:	Reservoir pressure	3000 psia
	Reservoir temperature	240° F
	Rock porosity	30 per cent
	Interstitial water	27 per cent
	Tank oil production	323 bbl per day
	Tank oil gravity	45° API
	Primary trap gas rate	3,765 MCF per day
	Primary trap gas gravity	0.65 (air = 1)
	Tank vapor rate	169.2 MCF per day
	Tank vapor gravity	1.25 (air = 1)

Solution: Basis: One acre-foot sand volume

Average separator gas gravity

$$= \frac{3{,}765 \cdot 0.65 + 169 \cdot 1.25}{3765 + 169} = 0.677$$

Producing gas-oil ratio

$$= \frac{(3765 + 169)(10^3)}{323} = 12{,}170 \text{ cu ft per bbl}$$

$$r \frac{(t)^{0.5}}{(\gamma_g)^{0.3}} (\delta_1)^{2.9 \times 10^{-0.00027 r}}$$

$$= 12170 \frac{(240)^{0.5}}{(0.677)^{0.3}} (0.8017)^{2.9 \times 10^{-0.00027 \cdot 12170}}$$

$$= 12170 \cdot \frac{15.5}{0.889} \cdot 0.999 = 2.125(10^5)$$

From Figure 39, formation volume at 3000 psia = 13.5 bbl per bbl of tank oil

One acre-foot reservoir rock volume has

$$43560 \cdot 0.30 \cdot (1 - 0.27) \cdot \frac{1}{5.615} = 1700 \text{ bbl of hydrocarbon space}$$

Therefore, tank oil volume $= \frac{1700}{13.5} = 127$ bbl per acre-foot

surface gas volume = 127 · 12170 = 1534 MCF per acre-foot

the correlation should be applied with caution to other fields.

The basis for the 160°F data, presented in Table 19, is shown in Figure 40. The curves illustrate a behavior of condensate systems, the understanding of which is important in the correct application of laboratory results to field conditions. The fact that the curves show maximum dew point pressure values makes it possible for a laboratory system to have a dew point pressure equal to the reservoir pressure, but not necessarily to have the same composition as the reservoir vapor. For example, the uppermost curve shown in Figure 40 exhibits a dew point pressure of 4700 psia over a gas-oil ratio range from 8,000 to 12,000 cu ft per bbl. If the producing gas-oil ratio at the time of sampling were 8000 cu ft per bbl and the reservoir pressure were 4700 psia, the data could be interpreted as indicating the reservoir fluid to be at its dew point and to have

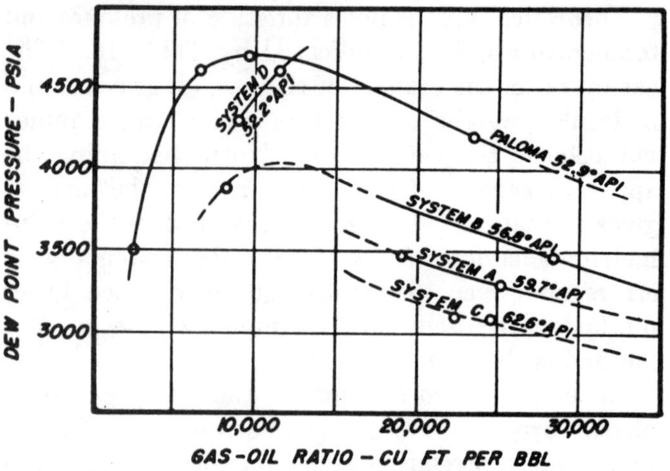

Figure 40—Influence of Gas–Oil Ratio and Tank Oil Gravity upon Retrograde Dew Point Pressure at 160°F (After Sage and Olds, Trans. AIME, 1947)

the properties of the 8000 cu ft per bbl system. On the other hand, it is not inconceivable that the actual reservoir vapor might be the equivalent of the 12,000 cu ft per bbl system and that the 8000 ratio was a fortuitous result of simultaneous production of vapor and retrograde reservoir liquid into the well bore. Under such circumstances, the use of the phase behavior of the 8000 cu ft per bbl system to predict the reservoir fluid behavior could result in errors in both the correct formation volume of the reservoir system and the quantity of retrograde liquid that would be formed as a result of reservoir pressure decline.

Because of the domelike shape of the gas-oil ratio-dew point pressure curve, one should not place as much emphasis on laboratory dew point pressure measurements as is usually placed on bubble point pressures. As will be shown in Chapter 7, there is a monotonic relation of bubble point pressure with gas-oil ratio, so that simultaneous flow of free gas and saturated liquid into the well bore will always cause the bubble point pressure of the recombined dissolved gas system to be higher than the bottom-hole pressure. On the other hand, simultaneous flow of retrograde reservoir liquid and saturated vapor into the well bore may result in a condensate system having a dew point pressure lower than the bottom-hole pressure and thus cause one to believe the reservoir fluid to be undersaturated.

Liquid Phase Behavior. The most characteristic curve of gas-condensate systems relates the quantity

Table 19. Relation of Dew Point Pressure of California Condensate Systems. Data of Sage and Olds[8]

1	2	3	4	5	6	7
Tank Oil Gravity	Gas—Oil Ratio, Cubic Feet per Barrel					
°API	15,000	20,000	25,000	30,000	35,000	40,000
			100°F			
52	4,440	4,140	3,880	3,680	3,530	3,420
54	4,190	3,920	3,710	3,540	3,410	3,310
56	3,970	3,730	3,540	3,390	3,280	3,180
58	3,720	3,540	3,380	3,250	3,140	3,060
60	3,460	3,340	3,220	3,100	3,010	2,930
62	3,290	3,190	3,070	2,970	2,880	2,800
64	3,080	3,010	2,920	2,840	2,770	2,700
			160°F			
52	4,760	4,530	4,270	4,060	3,890	3,650
54	4,400	4,170	3,950	3,760	3,610	3,490
56	4,090	3,890	3,690	3,520	3,380	3,270
58	3,840	3,650	3,470	3,320	3,200	3,110
60	3,610	3,430	3,280	3,150	3,040	2,960
62	3,390	3,240	3,100	2,990	2,890	2,810
64	3,190	3,060	2,930	2,820	2,740	2,670
			220°F			
54	4,410	4,230	4,050	3,890	3,750	3,620
56	3,990	3,780	3,600	3,440	3,300	3,180
58	3,700	3,480	3,280	3,110	2,970	2,850
60	3,430	3,210	3,030	2,880	2,760	2,660
62	3,150	2,970	2,800	2,670	2,570	2,480
64	2,900	2,740	2,590	2,470	2,380	2,300

of equilibrium liquid phase to pressure. This curve usually is determined at reservoir temperature to indicate the hydrocarbon liquid saturation that will be formed in the reservoir sands as a result of pressure decline. These data are important to economic studies of cycling operations. Three type curves are shown in Figure 41.

The curve illustrated by *A* shows the relation between volume per cent liquid and pressure for a flash process; the approach to the origin is merely an expression of the large total volume necessary to reach a low pressure for a system of constant over-all composition.

The curves illustrated by *B* and *C* are characteristic of differential processes in which the total volume of the system is maintained constant during the pressure reduction. Temperature and the initial composition of the system will affect the dew point pressure and the quantity of liquid at any pressure. In general, a system that has a surface producing a gas-oil ratio of 15,000 cu ft per bbl will give a maximum liquid content of 4 to 6 per cent by volume at reservoir temperatures near 200°F,

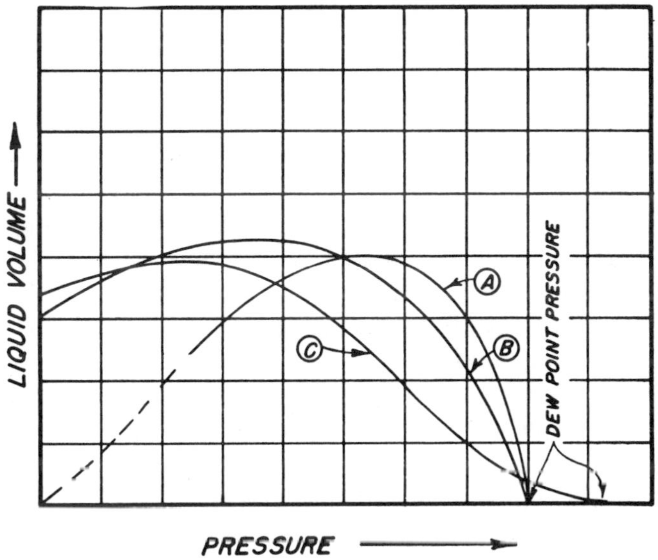

Figure 41—Curves Illustrating Several Types of Liquid Behavior of Condensate Systems

whereas a 50,000 cu ft per bbl system ordinarily will give less than 1 per cent liquid. Curves *B* and *C* illustrate differences in liquid volume behavior near the dew point pressure. While insufficient data are available in the literature to explain these differences, it is believed that the asymptotic approach to the dew point pressure of curve *C* is caused by a system that has a wider range of heavier components than a system that gives the *B* curve.

Where actual volumes of liquid formed during flash and differential processes have been compared, it has been found that the two liquid quantities have been nearly equal during the first 25 per cent reduction in pressure. As further pressure reductions occur, the differential process normally results in less liquid because portions of the heavier components are removed from the system as the differential process proceeds. In natural systems, the maximum amount of liquid occurs in the pressure range 1200 to 2500 psi for both the flash and differential processes. At lower temperatures larger amounts of liquid are formed and maximum liquid occurs at a lower pressure.

Vapor Phase Behavior. Just as the volume of liquid formed by a condensate system goes through a maximum near 2000 psi, the quantity of heavy components held in the vapor phase goes through a minimum, usually at about 1000 psia. This behavior of condensate systems is illustrated quite clearly by the results of the differential tests shown in Figure 42. The percentage of the system that has been produced is related to the resulting decline in system pressure, as shown by curve *A* of Figure 42. For example, 20 per cent of the system available in the vapor form, between the dew point pressure of 2960 and 14.7 psia, is produced during pressure decline from 2960 to 2370 psia. The relation presented in curve *A* can be used directly with field production data to calculate reserves where it is known that the reservoir hydrocarbon space does not change because of reservoir pressure decline. Curves of the type illustrated are characteristically S-shaped because of changes that occur in the compressibility factor and system composition as pressure declines.

Curve *B* of Figure 42 shows the amount of retrograde liquid formed during the differential test.

An indication of the richness of the vapor phase at any pressure is given by curve *C*, Figure 42. The horizontal line segments represent measured average gravities of the vapor removed from the pressure cell in the indicated pressure intervals. The smooth curve represents the instantaneous value of the vapor phase gravity. As will be shown in the next section, the calculated quantity of butanes plus

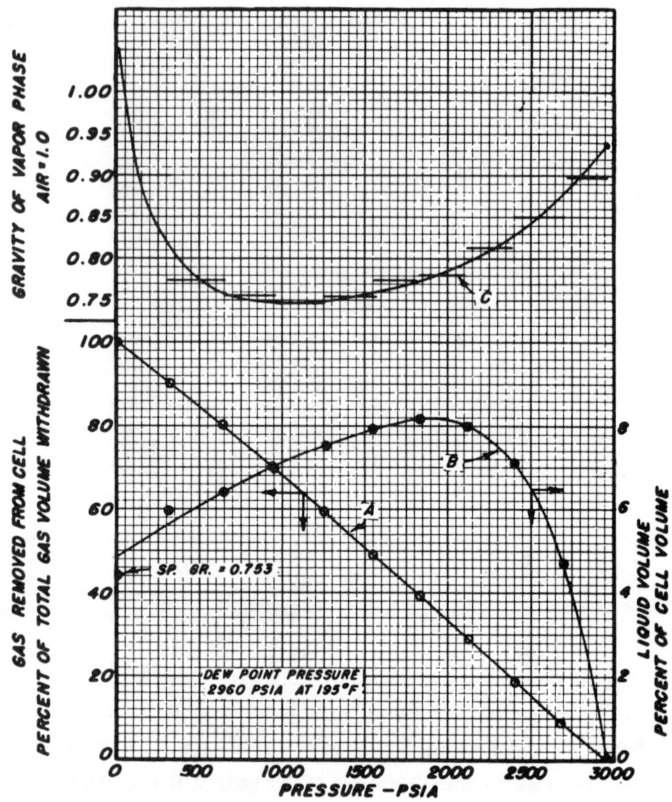

Figure 42—**Behavior of 2960 Psia Dew Point System at 195°F Caused by Differentially Removing Equilibrium Vapor**

material in the vapor phase decreases from 5.2 gal per MCF at the dew point to 2.6 gal per MCF at 1000 psia.

Phase Compositions

Laboratory tests on gas-condensate systems usually are made to determine the volumetric behavior of the systems at reservoir and surface trapping conditions. For special engineering studies, however, one may need to know the phase compositions at various pressures during the depletion of a condensate reservoir. The composition of each phase can be determined experimentally, but to do so requires more work than often is believed justified.

Methods of calculating phase compositions and volumes by use of equilibrium constants and the composition of the system already have been outlined in Chapters 3, 4, and 5. In gas-condensate systems, however, the calculated phase compositions and volumes, in particular the liquid phase volume, can be greatly in error through use of present known equilibrium constants. It becomes advisable, therefore, when making such calculations, to control them by use of experimentally determined volume measurements. The remaining portion of this chapter will discuss the steps required in such calculations. The example given pertains to an actual gas-condensate system that has a dew point pressure of 2960 psia at 195°F and the composition shown in Table 20. The results of laboratory differential tests to determine the volume of liquid phase and the pressure reduction-vapor withdrawal relation have been given previously in Figure 42.

As a starting point in the calculation, let it be assumed that a reservoir hydrocarbon space of 100 cu ft represents the unit working volume of the system in which we are interested. It will be assumed that this volume remains constant throughout the calculation process. The calculations in Table 20 determine that 54.0 moles constitute the original system at 2960 psia and 195°F. The calculations also show that 2.1 lb moles of material remain in the 100 cu ft volume at 14.7 psia and 195°F. The difference between this value and the original (51.9 lb moles) represents the quantity of material removed during the differential process.

The second step is to determine a set of equilibrium constants with which to calculate phase compositions. An immediate test of the applicability of any set of constants is that the sum of n_k/K_k must be equal to unity (Equation 31) at the observed dew point pressure of 2960 psia. In Table 21 are presented three dew point pressure calculations

TABLE 20. CALCULATION OF THE ORIGINAL AND FINAL QUANTITIES OF MATERIAL REMAINING IN THE UNIT RESERVOIR VOLUME

Basis: 100 cu ft hydrocarbon space
Initial dew point pressure 2960 psia
Final pressure 14.7 psia
Temperature 195°F

Initial condition at 2960 psia:

1	2	3	4	5	6
Component	Mole Fraction n	Critical Constant of Components Tc	Pc	Pseudo Critical Temperature and Pressure nTc	nPc
C_1	0 7527	344	673	258.7	506.0
C_2	0.0766	550	709	42.1	54.3
C_3	0.0441	666	618	29.4	27.2
C_4	0.0309	750	540	23.2	16.7
C_5	0.0221	838	484	18.5	10.7
C_6	0.0206	915	434	18.8	8.9
C_7+	0.0530*	1042	418	55.3	22.2
	1.0000			446.0	646.0

* C_7^+ molecular weight = 114
C_7^+ specific gravity = 0.755

$$pP_R = \frac{2960}{646.0} = 4.58 \qquad pT_R = \frac{655}{446.0} = 1.47$$

$$Z = 0.78 \text{ (Figure 15)}$$

$$N = \frac{PV}{ZRT} = \frac{2960 \cdot 100}{0.78 \cdot 10.73 \cdot 655} = 54.0 \text{ lb moles}$$

Final condition at 14.7 psia:

Specific gravity of cell residue = 0.753 60/60 (Fig. 42)
From Figure 25, density of cell residue at 14.7 psia and 195°F = 0.753 · 62.37 − 3.6 = 43.4 lb per cu ft
From Figure 38, assumed molecular weight of residue = 113
From Figure 42, volume of residue liquid at 14.7 psia and 195°F = 4.9 cu ft

$$\text{Quantity of residue liquid} = \frac{4.9 \cdot 43.4}{113} = 1.88 \text{ lb moles}$$

$$\text{Quantity of residue gas} = \frac{(100 - 4.9)}{379} \cdot \frac{520}{655} = 0.20 \text{ lb moles}$$

Total quantity of residue gas and liquid = 1.88 + 0.20 = 2.08 lb moles
Material removed from the unit volume in going from 2960 to 14.7 psia = 54.0 − 2.1 = 51.9 lb moles

TABLE 21. DETERMINATION OF RELATIVE EQUILIBRIUM CONSTANTS TO COMPUTE DEW POINT CONDITIONS

Basis: Roland, Smith, and Kaveler's Data 200°F

$$\sum_{k=1}^{k=j} \frac{n_k}{K_k} = 1$$

1	2	3	4	5	6	7	8
		Pressure 2960 PSIA		Pressure 3500 PSIA		Pressure 3600 PSIA	
Component	Mole Fraction n	K	n/K	K	n/K	K	n/K
C_1	0.7527	1.67	0.4500	1.35	0.5575	1.30	0.5795
C_2	0.0766	0.92	0.0830	0.96	0.0799	0.97	0.0790
C_3	0.0441	0.69	0.0649	0.82	0.0538	0.85	0.0519
C_4	0.0309	0.52	0.0594	0.71	0.0435	0.76	0.0407
C_5	0.0221	0.34	0.0650	0.54	0.0410	0.60	0.0379
C_6	0.0206	0.23	0.0896	0.44	0.0468	0.50	0.0412
C_7+	0.0530	0.095	0.5580	0.28	0.1892	0.35	0.1514
	1.0000		1.3690		1.0117		0.9816

Dew Point Pressure = 3540 psia

using Roland, Smith, and Kaveler's constants for 200°F. (It is not justified at this point to work out constants corrected to 195°F.) These constants satisfy the condition that the summation of n_k/K_k is equal to unity only at 3540 psia. As the system under study gave an observed dew point at 2960 psia, Roland's 3540 psia values must be used at 2960 psia pressure. Thus, the published constants are used here primarily for their relative values and not for their absolute values.

The methods presented in Chapter 5 are used next to estimate equilibrium constants at pressures below the dew point pressure. Assuming an apparent convergence pressure of the system several hundred psi above the dew point pressure makes it possible to calculate by use of Figures 31 and 32 the minimum K's and the pressure at which they occur. The smooth curves shown in Figure 43 have been drawn to connect the pressures at which the individual constants are equal to unity (vapor pressure), the minimum constants, the constants at the experimental dew point pressure, and a convergence pressure of 3500 psia. Roland, Smith, and Kaveler's curves at 200°F also have been plotted in Figure 43 to aid in drafting the 195°F curves. The curve constructed through the solid points will be explained later.

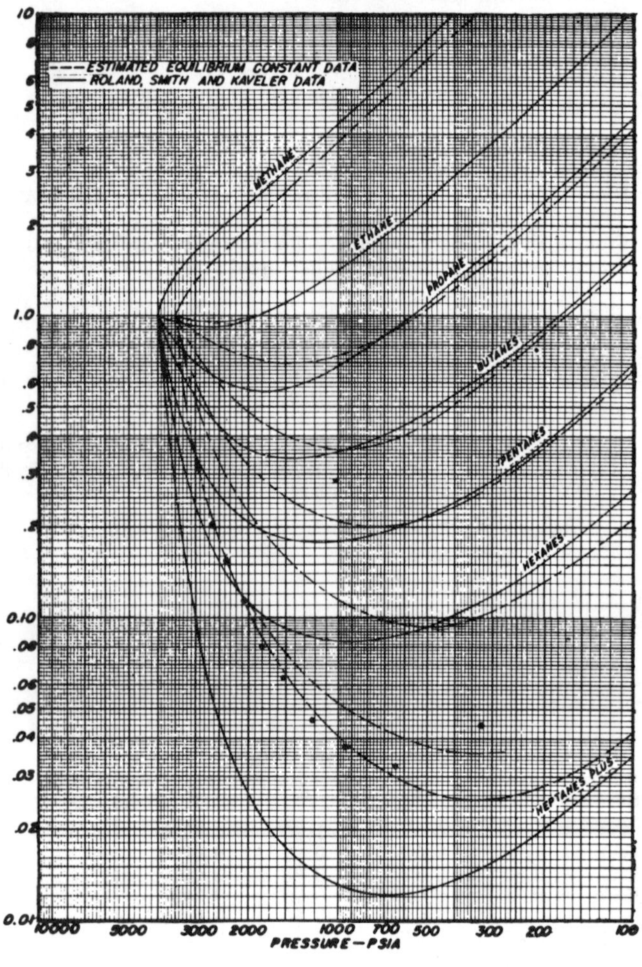

Figure 43—Estimated Equilibrium Constant Data at 195°F from Roland, Smith, and Kaveler Data at 200°F

The fourth step is to establish a calculation procedure to simulate the differential process of producing equilibrium vapor from the unit volume. A simple stepwise calculation process is diagrammed in Figure 44. This calculation process can be visualized

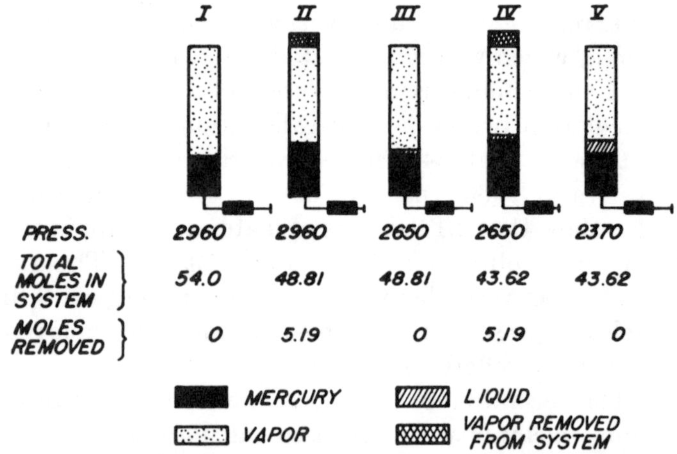

Figure 44—Diagram Illustrating Stepwise Calculation Process

best as taking place in a closed vessel in which it is possible to remove and inject an inert liquid such as mercury into the bottom of the vessel. Equilibrium vapor phase material can be removed from the top of the vessel. Referring to Figure 44, condition I pictures 54.0 lb moles of original dew point material at 2960 psia. Condition II pictures the removal of 5.19 moles of vapor from the vessel at a constant pressure of 2960 psia, leaving a total of 48.81 lb moles of material in the system at condition III. In going from condition II to condition III, the exact amount of mercury previously injected into the vessel is withdrawn, so that the total volume of the system at condition III again is 100 cu ft. The volume that the liquid phase occupies at 2650 psia is read from curve B, Figure 42. At 2650 psia, a second unit of 5.19 moles of equilibrium vapor is removed from the vessel. The second modified system pictured by IV then is brought to the original 100 cu ft volume at 2370 psia, as pictured by condition V. The process then is repeated at pressures of 2090, 1800, 1520, 1230, 940, 640, 320, and 15 psia. These particular pressures are read from curve *A*, Figure 42, and represent the pressures which result from withdrawal of ten per cent increments, or 5.19 lb moles of equilibrium vapor phase.

An alternate process can be pictured in which the 5.19 lb moles are removed from the vessel after pressure reduction, rather than before. The composition of the vapor and liquid phases calculated by this second process does not differ materially from that calculated by the process just described. Of course, the greater the number of steps, the closer the two methods will agree with each other and, for an infinite number of steps, each process becomes a true differential process.

The PVT data presented in Figure 42 are in terms of volumes, whereas equilibrium constant calculations make use of molal phase quantities. To convert the cubic feet of liquid phase calculated from Figure 42 to pound moles of liquid, it is necessary to know the molecular weight and density of the liquids as they exist at the elevated pressures and 195°F. These values can be determined by the methods presented in Chapter 4, provided the pressure-density and pressure-molecular weight relations of the heptanes plus fraction are known. The only direct laboratory data available are the molecular weight and specific gravity of the total heptanes plus in the original system and the specific gravity of the residual liquid remaining in the laboratory cell at the completion of the differential process. These data have been used with the trend of the curves in Figure 29 to prepare the assumed pressure relations shown in Figure 45. The measured values of 114 and 0.755 are shown as corresponding to the dew point vapor. Inasmuch as the residual cell liquid at 14.7 psia and 195°F is of the order of 85 per cent heptanes plus material, its measured specific gravity of 0.753 has been corrected upward to correspond to a heptanes plus specific gravity of 0.760. From this gravity, the molecular weight of the heptanes plus in the residue liquid is estimated to be 117. Unfortunately, these are all the data available from which to draw the curves shown in Figure 45. Whether the curves are straight lines or have curvature as shown can only be surmised. The heptanes plus liquid density at 2960 psia is an assumed value, because the density cannot be calculated from the existing data. In short, Figure 45 illustrates one of the great needs

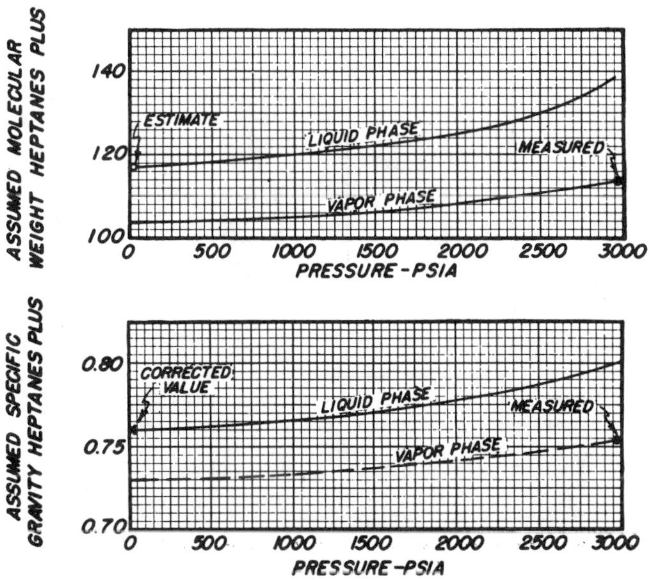

Figure 45—Assumed Relationship of Molecular Weight and Specific Gravity of Heptanes Plus to Pressure

in vaporization calculation work: the behavior of the heptanes plus fraction at different pressures.

The composition of the dew point liquid at 2960 psia and 195°F is shown in Table 22. The method of obtaining the equilibrium constants listed in column 3 has been discussed previously in connection with Figure 43. The properties of the heptanes plus fraction required to calculate its volume and weight were obtained from Figure 45. Volumes of the individual components listed in column 6 were obtained by use of the atmospheric liquid densities of Table 1. An arithmetic average of the liquid densities of iso- and normal butane and pentane was used in the calculation. It is interesting to note that almost 20 weight per cent of the dew point liquid is methane.

The calculation of vapor and liquid compositions at 2650 psia is shown in Table 23. In part B the number of moles of liquid and vapor in equilibrium is calculated from the total moles (48.81) comprising the first modified system and the quantity of liquid phase determined by the laboratory tests. To convert the measured liquid volume to pound moles, it is assumed that the factor, pound moles per cubic foot, evaluated at 2960 psia (Table 22) also will apply at 2650 psia. This assumption is checked

(5) The mole fractions of the components in the vapor phase are made to add to unity by minor adjustments in the mole fraction of the methane.

TABLE 22. CALCULATION OF PROPERTIES OF DEW POINT LIQUID AT 2960 PSIA AND 195°F
Basis: One pound mole of liquid

1	2	3	4	5	6
Component	System Composition Mole Fraction n	Equilibrium Constant at 2960 PSIA & 195°F K	Liquid Phase Composition $\frac{n}{K}$	Weight of Components lb per lb mole liquid $\frac{nM}{K}$	Liquid Volume of Components at 14.7 PSIA & 60°F cu ft $\frac{nM}{K\rho}$
C_1	0.7527	1.32	0.5720	9.15	
C_2	0.0766	0.96	0.0799	2.40	
C_3	0.0441	0.83	0.0531	2.34	0.074
C_4	0.0309	0.74	0.0418	2.43	0.068
C_5	0.0221	0.57	0.0388	2.80	0.071
C_6	0.0206	0.47	0.0438	3.77	0.091
C_7+	0.0530	0.31	0.1706	23.55*	0.472*
	1.0000		1.0000	46.44	0.776

* Assumed molecular weight of C_7^+ = 138 (Figure 45)
Assumed specific gravity of C_7^+ = 0.800 (Figure 45)

Density propane plus at 14.7 psia and 60°F $= \frac{35.45}{0.776} = 45.7$ lb per cu ft

Weight per cent ethane in ethane plus $= \frac{2.40}{37.63} = 6.3$ per cent

Weight per cent methane in system $= \frac{9.15}{46.44} = 19.5$ per cent

From Figure 23, pseudo-liquid density at 14.7 psia and 60°F = 31.3 lb per cu ft
From Figure 24, density at 2960 psia and 60°F = 31.3 + 2.7 = 34.0 lb per cu ft
From Figure 25, density at 2960 psia and 195°F = 34.0 − 9.3 = 24.7 lb per cu ft

Pound moles liquid per cubic foot liquid $= \frac{24.7}{46.44} = 0.532$

and, if necessary, is corrected after the composition of 2650 psia liquid has been calculated in part B.

Part B of Table 23 outlines the vaporization calculation at 2650 psia, using the relative values of L and V determined in part A. The following procedure has been followed in making these calculations:

(1) With the exception of heptanes plus, the mole fraction of each component in the liquid phase is calculated by the usual methods.

(2) The mole fraction of heptanes plus in the liquid phase (0.2059) is set at the value necessary for $\Sigma x = 1$.

(3) Working backward from the values of x_7^+ and n_7^+, the values of $L + VK_7$ are calculated.

(4) The mole fraction of heptanes plus in the vapor phase, y_7^+, is calculated from the values of K_7^+ and x_7^+, using the relation $y = Kx$.

The procedure just outlined assumes that all equilibrium constants are correct, except that of the heptanes plus fraction. This assumption, of course, may not be strictly valid, but it presents a workable method of calculating the phase compositions. The net effect of throwing all equilibrium constant errors into the heptanes plus constant is shown by the solid points in Figure 43. If the methane constant curve had been drawn slightly higher, so that its value at 2650 psia was 1.60 rather than 1.48, the calculated equilibrium constant of the heptanes plus fraction would have been 0.173 rather than 0.217. This would have caused better agreement between the calculated heptanes plus curve and the curve previously obtained by consideration of the apparent convergence pressure. However, unless a much wider discrepancy is obtained than that illustrated in Figure 43, it usually does not pay to undertake further refinement of the curve.

The calculations in part C of Table 23 are required to check the assumed value of 0.53 lb moles per cu ft of liquid used in part A. The agreement of 0.53 with **0.536** is considered satisfactory. Had a check not been obtained, it would have been necessary to recalculate parts A, B, and C.

The calculations in part D of Table 23 correspond to the removal of 5.19 lb moles of 2650 psia equilibrium vapor, as pictured by condition IV of Figure 44. The remaining moles of material constitute the second modified system that will be expanded to 2370 psia.

The results of the ten-step calculation process are shown graphically in Figures 46 and 47. The composition, density, and molecular weight of the liquid phase at various pressures are shown in Figure 46. As might be expected, the quantity of dissolved methane continuously decreases as the pressure on

TABLE 23. CALCULATION OF VAPOR AND LIQUID COMPOSITION AT 2650 PSIA

Part A: Estimation of Vapor to Liquid Ratio

Basis: 48.81 lb moles of system

From calculations in Table 22, assume that the liquid phase will have the equivalent of 0.53 lb moles per cu ft.
From Figure 42, volume of liquid at 2650 psia is 5.2 cu ft. Therefore, moles of liquid = 5.2 · 0.53 = 2.76.

Mole fraction liquid, $L = \frac{2.76}{48.8} = 0.056$: Mole fraction vapor, $V = 0.944$

Part B: Calculation of Vapor and Liquid Composition

Basis: One pound mole of system

1	2	3	4	5	6	7
	System Composition Mole Fraction	Equilibrium Constant at 2650 Psia & 195°F	L = 0.056 V = 0.944		Liquid Phase Composition Mole Fraction	Vapor Phase Composition Mole Fraction
Component	n	K	VK	L + VK	x	y
C_1	0.7527	1.48	1.397	1.453	0.5180	(0.7659)
C_2	0.0766	0.95	0.897	0.953	0.0804	0.0764
C_3	0.0441	0.82	0.774	0.830	0.0531	0.0435
C_4	0.0309	0.63	0.595	0.651	0.0475	0.0299
C_5	0.0221	0.47	0.444	0.500	0.0442	0.0208
C_6	0.0206	0.37	0.349	0.405	0.0509	0.0188
C_7^+	0.0530	(0.217)	(0.205)	(0.261)	(0.2059)	0.0447
	1.0000				1.0000	1.0000

Part C: Properties of Liquid Phase

Basis: One pound mole of liquid

1	2	3	4
Component	Liquid Phase Composition Mole Fraction x	Weight of Components lb xM	Liquid Volume of Components at 14.7 psia & 60°F cu ft $\frac{xM}{\rho}$
C_1	0.5180	8.31	
C_2	0.0804	2.42	
C_3	0.0531	2.34	0.074
C_4	0.0475	2.76	0.076
C_5	0.0442	3.19	0.082
C_6	0.0509	4.39	0.106
C_7^+	0.2059	27.18*	0.552
	1.0000	50.59	0.890

* From Figure 45, molecular weight $C_7^+ = 132$
specific gravity $C_7^+ = 0.790$

Density propane plus at 14.7 psia and 60°F $= \frac{39.86}{0.890} = 44.8$ lb per cu ft

Weight per cent ethane in ethane plus $= \frac{2.42}{42.28} = 5.7$

Weight per cent methane in system $= \frac{8.31}{50.59} = 16.4$

From Figure 23, pseudo-liquid density at 14.7 psia and 60°F = 33.2 lb per cu ft

From Figure 24, density at 2650 psia and 60°F = 33.2 + 2.0 = 35.2 lb per cu ft

From Figure 25, density at 2650 psia and 195°F = 35.2 − 7.9 = 27.3 lb per cu ft

Pound moles liquid per cu ft liquid $= \frac{27.3}{50.59} = 0.539$

(Assumed value of 0.53 in Part A is satisfactory.)

Table 23—(*Continued*)

Part D: Calculation of Modified System at 2370 Psia
Basis: 48.81 lb moles of material at 2650 psia
5.19 lb moles of 2650 psia vapor removed

1	2	3	4	5	6	7
	First Modified System				Second Modified System	
	2650 Psia System		2650 Psia Vapor		2370 Psia System	
Component	Mole Fraction	Pound Moles	Mole Fraction	Pound Moles	Pound Moles	Mole Fraction
C_1	0.7527	36.740	0.7659	3.975	32.765	0.7511
C_2	0.0766	3.739	0.0764	0.396	3.343	0.0766
C_3	0.0441	2.152	0.0435	0.226	1.926	0.0442
C_4	0.0309	1.508	0.0299	0.155	1.353	0.0310
C_5	0.0221	1.079	0.0208	0.108	0.971	0.0223
C_6	0.0206	1.005	0.0188	0.098	0.907	0.0208
C_7^+	0.0530	2.587	0.0447	0.232	2.355	0.0540
	1.0000	48.810	1.0000	5.190	43.620	1.0000

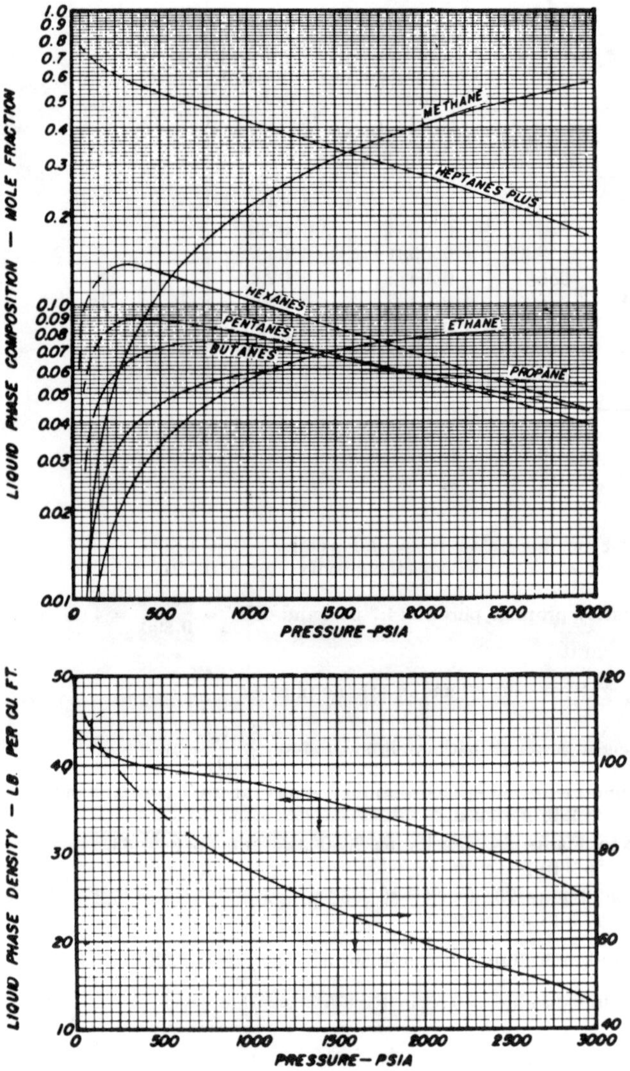

Figure 46—Calculated Composition and Properties of Liquid Phase

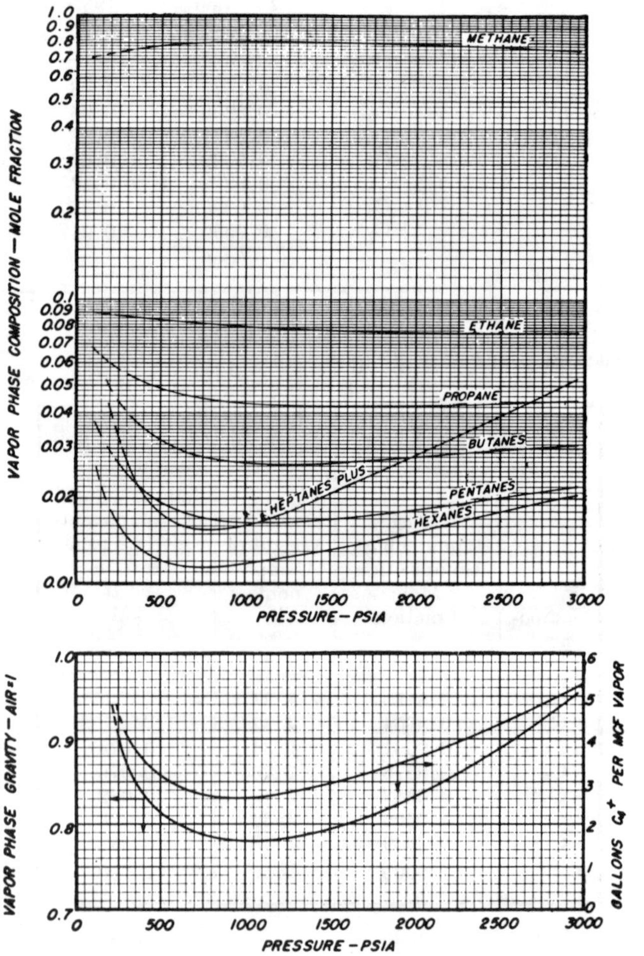

Figure 47—Calculated Composition and Properties of Vapor Phase

the system is reduced. Likewise, the proportion of heavy components increases with decreasing pressure. Intermediate components tend to have maximum concentration in the liquid phase at pressures near the pressure of their minimum equilibrium constant. For example, the maximum butane concentration in the liquid is near 850 psia, and the equilibrium constant of butanes (Figure 43) is a minimum at that pressure. The behavior of the vapor phase is shown in Figure 47. It will be noted that the calculated minimum gravity of 0.780 at 1000 psia does not check the laboratory test value of 0.75 shown in Figure 42. This may be caused in part by the high value of the heptanes plus equilibrium constant. A second factor that may contribute to the discrepancy is the assumed molecular weight of the heptanes plus vapor.

It can be appreciated from the foregoing section that calculated compositions of the vapor and liquid phases in a gas-condensate system may be in error because of any one of the numerous assumptions that have to be made in the calculation process. The method outlined above, where the volumetric behavior of the system determined experimentally in the laboratory is used as a control in the calculation process, is believed to be the most satisfactory one available. The accuracy of these calculations can be expected to improve as additional data become available on the behavior of condensate systems.

Summary of Principal Points

Gas-condensate systems are identified by their high gas-oil ratios and lightly colored, high-gravity tank oils. The critical temperature of the naturally occurring condensate systems is lower than the temperature of the reservoir in which they are found. The systems normally exhibit retrograde behavior at pressures above 2000 psia because of the relation of the critical temperature of the system to the reservoir temperature.

The volumetric behavior of gas-condensate systems can be calculated easily by using the pseudo-critical constant method proposed by W. B. Kay.[12] In many instances the pseudo-critical constants of a series of gas-condensate systems can be plotted against the gravity of the system; calculation charts then can be constructed that will give the volume of the system as a function of pressure, temperature, and system gravity. Figure 38 presents an easy way of calculating the gravity of a well fluid from the quantities and gravities of the resulting surface gases and tank oil.

Naturally occurring gas-condensate systems, in general, contain greater than 75 mole per cent methane. The volume of the liquid phase in equilibrium with vapor under reservoir conditions seldom is greater than 10 per cent of the total volume occupied by the system. At pressures of the order of 3000 psia or greater, the retrograde liquid in a reservoir may contain up to 40 per cent liquid methane by volume. The density of the liquid under these conditions often is as low as 25 lb per cu ft.

The quantity of butane plus material held in the vapor phase usually is at a minimum near 1000 psia. This means that production from high pressure condensate fields should follow an increasing gas-oil ratio trend until the reservoir pressure declines to near 1000 psia. The producing gas-oil ratio normally will decrease at pressures less than 1000 psia.

Calculation of phase compositions is made best by using experimentally determined phase volumes to control the equilibrium calculations. However, it should be recognized that the calculated compositions can be in error because of using nonrepresentative equilibrium constants and incorrect assumptions regarding the pressure-density and pressure-molecular weight relations of the heptanes plus fractions.

CHAPTER 7

DISSOLVED GAS SYSTEMS

Most naturally occurring hydrocarbon systems behave in a manner that involves bubble points. In terms of multi-component system behavior discussed in Chapter 1, such a system has a critical temperature greater than the temperature at which it exists in a reservoir. To have a critical temperature of 200°F or higher, it is necessary for the system to contain a large amount of the heavier hydrocarbons. Such systems, when produced through surface trap settings, will give gas-oil ratios less than 2500 cu ft per bbl.* The tank oil normally is deeply colored, and the gravity may be as high as 45°API.

Pressure decline in dissolved gas-type reservoirs causes a gas phase to be formed within the reservoir rock pores. The mechanism of oil production caused by gas coming out of solution and expanding is termed a "dissolved" gas drive, or a "solution" gas drive, and usually is thought of in connection with reservoirs having no original free gas phase. A related producing mechanism exists in the oil zone of a reservoir initially containing free gas phase accumulated in a gas cap. While the pressure-production behavior of a reservoir under a dissolved gas drive is different from that resulting from an expanding gas cap, the thermodynamic behavior of the reservoir liquid phase during pressure decline is the same under the two mechanisms.

Laboratory tests to determine the PVT, gas solubility, and viscometric behavior of dissolved gas systems are standardized to a greater degree than tests performed on gas-condensate systems. Data on the compressibility and thermal expansion of the system, the quantity of gas in solution in oil at various pressures and temperatures, the shrinkage of the liquid hydrocarbon phase when gas is removed from solution under both reservoir and surface trapping conditions are determined routinely in many laboratories.

* The indicated 2500 cu ft per bbl solution gas-oil ratio represents approximately the maximum amount of gas that can be dissolved in oil at pressures up to about 10000 psi.

This chapter will describe first the apparatus and methods used in making laboratory tests. Data characteristic of the behavior of dissolved gas systems under surface trapping conditions will be presented in the second section of the chapter. The third section deals with the behavior of dissolved gas systems at reservoir pressures and temperatures. The last section will describe methods of adjusting the laboratory data to apply more fully to field conditions.

Laboratory Tests

Apparatus. The essential pieces of laboratory equipment consist of a pressure cell in which equilibrium can be obtained between gas and liquid phases, a controlled temperature bath with a range from 60 to 300°F, a pressure-measuring system, and a calibrated pump for injecting mercury into the cell in order to change the volume occupied by the hydrocarbon system.

Many different forms of pressure cells have been used. Acceptable work has been done with a cell constructed from two bull plugs welded together and suspended in a water bath. While this describes the crudest type of cell, it has the basic features of many cells in use today. The main operational feature of this type of cell is that only the total volume of the hydrocarbon system in the cell can be determined. When two phases are present, the volume of either phase cannot be determined separately without removing the other phase from the cell. For discussion purposes, a cell like this will be called a Type I cell. An illustration of such a cell is in Figure 48. More elaborate pressure cells

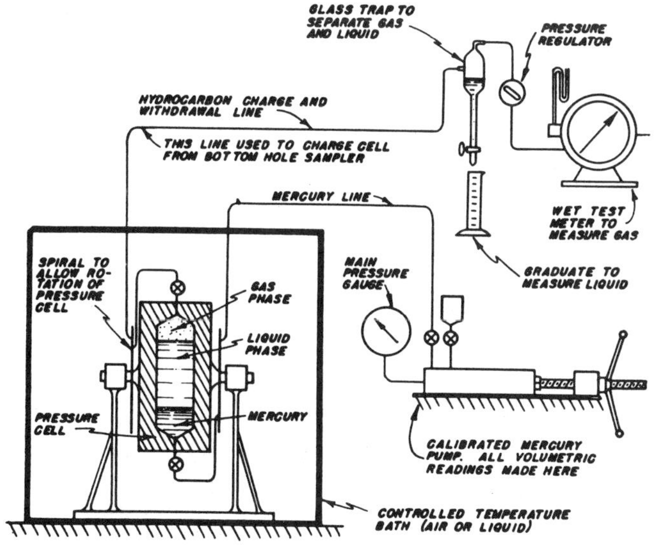

Figure 48—Sketch Illustrating Main Features of Type I PVT Cell

at California Institute of Technology[38] and at Calfornia Research Corporation permit the simultaneous determination of multiple-phase volumes. These are called Type II cells. A visual, or Type III, cell is one in which the position of the gas-oil and oil-mercury interferes within the cell is observed through glass windows.

In the Type I cell, the hydrocarbon volume is not measured directly, but is calculated as the difference between the total volume of the cell and the quantity of mercury within the cell. The quantity of mercury in the cell at any pressure is determined by keeping a running tally of the mercury injected into or withdrawn from the cell by means of a calibrated injection pump. Several corrections must be applied to the injection pump volume readings when changing from one pressure to another. For example, before it represents the change in volume of the hydrocarbon system within the cell, the volume of mercury injected or withdrawn by the pump must be corrected for the expansion of the cell itself, the compressibility of the mercury within the cell, and the change in volume of the metered mercury due to the thermal expansion of the increment being forced into the cell.

In the Type II cell illustrated in Figure 49, the total volume of the hydrocarbon system is determined by locating the mercury-hydrocarbon liquid interface within the cell with an electrically energized needle. The interface detector head is

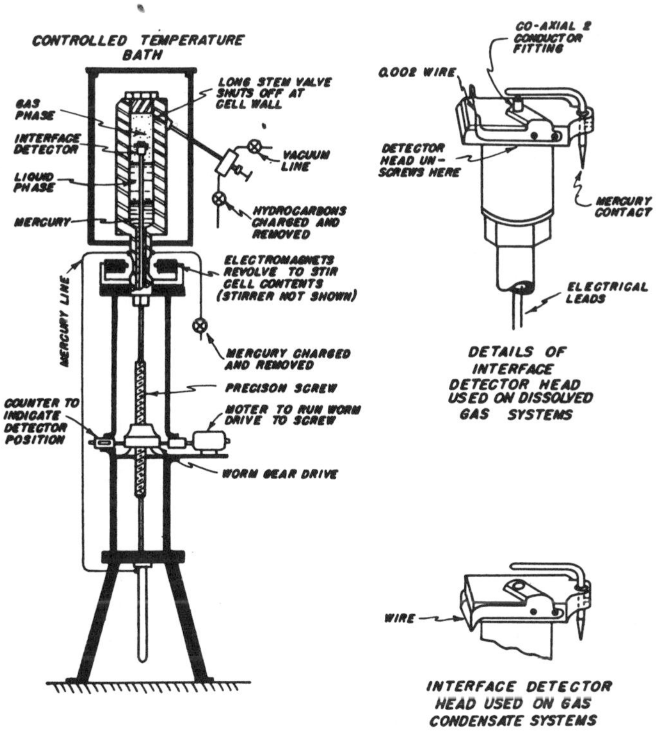

Figure 49—Sketch Illustrating Main Features of Type II PVT Cell

carried on a traveling rod which is coupled to a counting mechanism to indicate the precise location of the detector within the cell. The counter reading, when the mercury contact just touches the mercury surface in the cell, can be converted into the total volume of the hydrocarbon system by previous calibration.

In addition to the mercury contact, the traveling rod also carries a short length of 0.002 in. diameter platinum wire mounted in a horizontal plane. By forcing a small electrical current (200 milliamperes) through the wire and measuring the resulting voltage drop across the wire, a measure of the rate of heat transfer from the wire is obtained; i.e., the resistance of the wire is a function of its temperature, which in turn is a function of the rate at which heat is dissipated from the wire. As the heat transfer is different in vapor than in liquid, the wire can be used to locate the position of the vapor-liquid interface within the cell. When the method just described is used, the California Research

CALIFORNIA RESEARCH CORPORATION
LA HABRA LABORATORY

Test No. 5509
Sheet No. 1
Refer Book No. 226

P-V-T TEST DATA

Observers DYKSTRA
Date APRIL 30, 1945

Sample Bomb Nos.
Liquid 57 Gas
BOTTOM HOLE SAMPLE

Field DOUGLAS
Well 423-24A

Calibration:
V_t 0.04883 N_2 - 207.22 V_L 0.04883 ON + 32.67

$P = AS + C.B. + KN_2 - E$
A 1.0012 C.B. 14.6 K — E 36.2

Cell Temp. °F T_s	Stirring Time On	Stirring Time Off	~~VOLUMES~~ Total Screw N_2	Total Ml. V_t	Liquid Screw N_1	Liquid Difference N_2-N_1	Liquid Ml. V_L	% Liquid V_L/V_t	PRESSURE Obs. P.S.I. S	KN_2	Abs P.S.I.A. P	REMARKS	Formation Volumes Total V_{FT}	Formation Volumes Liquid V_{FL}
BUBBLE	POINT		AT	185°F					K =	4.837	× 10^{-4}			
185.3	0830	0920	6012.5	86.37					3670	2.9	3655			
			6022.3	86.85					3309	✓	3294			
			6032.0	87.52					2965	✓	2950			
			6041.0	87.76					2635	✓	2620	B.P. 'Y'		
185.2	0935		6048.7	88.14					2516	✓	2501	2520 PSIA 87.92 CC		
	0944		6058.0	88.59					2479	✓	2463	3.0391		
185.2	0953		6069.4	89.15					2437	✓	2421	2.9249		
	1003		6083.0	89.81					2386	✓	2370	29451		
	1014		6102.7	90.77					2325	3.0	2309	2.8195		
	1033		6123.1	91.77					2263	✓	2247	2.7750		
			6153.6	93.26					2178	✓	2162	2.7265		
185.3	1051		6194.0	95.23					2077	✓	2061	2.6786		
	1104		6234.2	97.20					1992	✓	1976	2.6085		
	1115		6286.8	99.76					1892	✓	1876	2.5492		
	1126		6342.1	102.46					1803	✓	1787	2.4804		
	1152		6394.4	105.02	5392.4	1002.0	81.60		1726	✓	1710	2.4355		
185.3			6503.8	110.36	5523.0	980.8	80.56		1589	3.1	1572	2.3629		
	1237		6723.8	121.10	5775.6	948.2	78.97		1378	3.3	1361	2.2566		
	1251		7022.7	135.70	6102.9	919.8	77.58		1165	3.4	1148	21991		
	1306		7586.3	163.29	6709.0	877.3	75.51		943	3.7	926	2.0081		
	1325		8467.9	206.27	7627.1	840.8	73.73		725	4.1	708	1.9013		
185.3	1340		9819.2	272.25	9017.6	801.6	71.81		548	4.8	532	1.7824		
	1400		11070.3	333.34	10291.5	778.8	70.70		450	4.5	434	1.7219		
	1420		12441.6	400.30	11680.4	761.2	69.84		377	6.0	362	1.6778		
TEMP.	CHANGE	TO	165°F						K =	4.847	× 10^{-4}			
165.0	1440	1500	5990.0	85.27					3686	2.9	3671			
165.2			6002.1	85.86					3187	✓	3172			
165.2			6010.3	86.26					2890	✓	2875			
165.1			6019.4	86.71					2553	✓	2538	B.P.		
165.1			6026.0	87.03					2427	✓	2411	2437 PSIA 86.88CC		

Table 24—Pressure-Volume-Temperature Test Data—Douglas Saturated Oil

Corporation apparatus is capable of measuring the total volume of hydrocarbon systems in the range of 30 to 450 cc with a precision of ±0.05 cc. Liquid phase measurements are equally precise but have a lower limit of 1 cc.

Flash Vaporization Tests. In a previous chapter, a flash vaporization was defined as a constant-composition variable volume process. Flash vaporization tests are carried out on dissolved gas systems to determine how the system behaves both in the reservoir and at the surface. To assist the reader to obtain a clear understanding of laboratory flash vaporization tests, this and the next chapter will discuss data determined on a bottom hole sample saturated at 2520 psia at 185°F. This particular system will be referred to as Douglas saturated oil.

A test sheet showing the laboratory data taken in evaluating the pressure-volume relation of Douglas saturated oil at 185°F is presented in Table 24. The pressure-volume readings, with similar readings at other temperatures, are plotted in Figure 50.

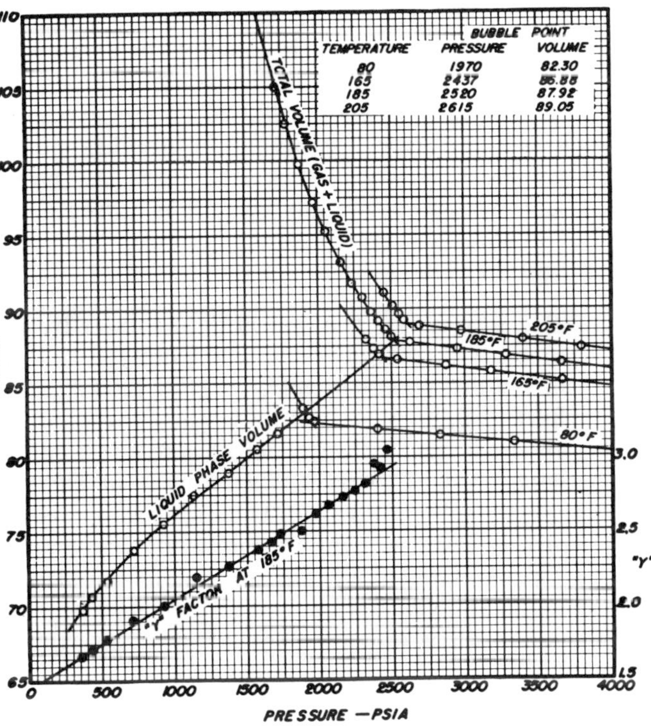

Figure 50 Pressure Volume Temperature Relationship of Douglas Saturated Oil

The pressure and volume of the system at the bubble point are evaluated easily by the "break" in the P-V curves.

The data presented in Figure 50 are the basic volumetric data required to define several properties of the system under reservoir conditions. The compressibility and thermal expansion of the system can be evaluated, as well as the volume that the system occupies at any pressure relative to its volume at some reference condition. For example, the compressibility of the system, β, at 3000 psia and 185°F can be calculated by using the definition of compressibility given in the following equation:

$$\beta = -\frac{1}{V}\left(\frac{\delta V}{\delta P}\right)_T \tag{35}$$

Substituting values from Figure 50, using the 2750 and 3750 psia values to define $\left(\frac{\partial V}{\partial P}\right)_T$,

$$\beta = -\frac{1}{87.25}\left(\frac{87.55 - 86.20}{2750 - 3750}\right) = 15.5(10^{-6}) \text{ vol/vol/psi}$$

The thermal expansion, α, at 3000 psia and 185°F can be obtained by cross-plotting the temperature-volume values at 3000 psia and applying the equation:

$$\alpha = \frac{1}{V}\left(\frac{\delta V}{\delta T}\right)_P \tag{36}$$

Over the small temperature range between 165°F and 205°F, $\left(\frac{\partial V}{\partial P}\right)_P$ is very close to being linear with temperature so that the thermal expansion of the system is calculated as follows:

$$\alpha = \frac{1}{87.25}\left(\frac{88.50 - 86.05}{205 - 165}\right) = 7.02(10^{-4}) \text{ vol/vol/°F}$$

If the volume that the system occupies at its bubble point is used as a reference state, the volume of the system at any other state can be expressed by a simple ratio. For example, the total volume of the system studied in the laboratory at 2000 psia and 185°F is shown in Figure 50 to be 96.6 cc. Referring this volume to the bubble point volume of 87.92 cc at 2520 psia and 185°F gives a relative volume of $\frac{96.6}{87.92} = 1.099$. Use of relative volumes has the advantage that the values are independent of the extent of the system, i.e., the laboratory system having a relative volume of 1.099 cc per cc of saturated oil is analogous to a reservoir oil system having a relative volume of 1.099 bbl per bbl of saturated oil.

The volume of most dissolved gas systems at pressures greater than their bubble point pressures varies linearly with pressure. As a consequence, it usually is not difficult to evaluate with good precision the change in volume of the system caused by a pressure change of a few psi. At pressures less than the bubble point pressure, the P-V relation of the system takes a form similar to that of a hyperbola. Under these conditions, there are several ways of smoothing the experimental data in order to evaluate the small system volume changes often required in material balance calculations. Probably the best method of smoothing P-V data at pressures below the bubble point pressure is by use of an empirical relation termed a "Y" correlation. This empirical relation can be expressed as follows:

$$Y = \left(\frac{P_B - P}{P}\right) \Big/ \left(\frac{V}{V_B} - 1\right) \tag{37}$$

where
P_B = bubble point pressure (psia)
P = pressure at any point
V_B = bubble point volume, consistent units
V = two-phase volume at pressure P, consistent units
Y = empirical value

Systems composed almost entirely of hydrocarbons show a linear relation of Y and pressure. This fact is illustrated by the lower curve in Figure 50. Systems that contain appreciable quantities of nonhydrocarbons usually give a curved relation of Y and pressure. The sensitiveness of the Y correlation is greatest close to the bubble point of the system. An example is the plotted point at the right end of the curve shown in Figure 50. To bring this 2463 psia point onto the line requires a correction in the system volume from 88.59 cc to 88.62 cc, a change of less than 0.04 per cent of the volume reading.

The properties of a fluid being produced from a well are evaluated from results of laboratory tests by reference to the system's bubble point pressure or gas-oil ratio. Thus, one often speaks of "the compressibility of the 2520 psia saturated oil" or "the formation volume factor of the 525 cu ft per bbl system." In the first instance, the term "saturated oil" is used synonymously with "bubble point," and it is assumed that one is referring to a particular temperature at which the fluid is at its bubble point conditions at 2520 psia. In the second instance, it is presupposed by the use of a gas-oil ratio value that one knows the method whereby the gas and oil are separated. As the gas-oil ratio of a system is affected by the pressures, temperatures, and number of equilibrium steps used to effect the separation, it is necessary to define explicitly the process by which the gas-oil ratio value was obtained, as well as the particular gas-oil ratio value.

The laboratory method of determining the gas-oil ratio of a particular system under surface trapping conditions is essentially as follows:

(1) The system is brought to equilibrium at the pressure and temperature condition of the primary trap in the field and the volumes of the liquid and gas phases measured. The resulting gas and liquid in the laboratory cell are comparable to the fluids leaving the primary trap, provided that thermodynamic equilibrium is established in the trap.

(2) The equilibrium gas phase is removed from the top of the cell, maintaining the cell pressure constant during the process by injecting mercury into the bottom of the cell. The volume of the gas phase is determined by collecting the gas in previously evacuated bottles and noting the change in pressure within the bottle. By means of the gas law equation, the observed gas volume is corrected to the standard reference condition of 14.7 psia and 60°F.

(3) The gravity of the gas is determined by removing a portion of the gas from the bottles into a previously evacuated glass flask. This flask is light in weight and can be weighed on an analytical balance before and after filling with gas. The molecular weight of the gas can be calculated from the increase in weight of the flask per unit change in flask pressure. This is shown by the following equations: From the gas law,

$$PV = NZRT \tag{38}$$

$$N = \frac{m}{M} \tag{39}$$

where
m = weight of gas of molecular weight M

Therefore substituting and solving for M,

$$M = \frac{mZRT}{PV} \tag{40}$$

Letting Δm and ΔP indicate the change in weight and pressure of the previously evacuated flask,

$$M = \frac{\Delta m}{\Delta P} \cdot C \tag{41}$$

In Equation 41, C is a constant which is primarily a function of temperature (the volume of the flask is constant; Z does not vary appreciably from unity, as the measurements are always carried out at less than atmospheric pressure). The factor C can be determined for any particular flask by going through the same process, but substituting air for gas.

(4) The liquid remaining in the cell from the first separation is brought to equilibrium at the pressure and temperature of the second-stage trap, and the procedures outlined above in 2 and 3 are repeated. The procedures again are repeated at the stock tank pressure and temperature.

(5) The volume of liquid remaining in the cell (tank oil) is measured and its volume corrected to 60°F by use of the National Petroleum Tables.[23] A portion of the liquid is removed and its density determined.

The laboratory data obtained while determining the surface gas-oil ratio of the Douglas sample under a two-stage separation process are shown in Table 25. Part A of Table 25 lists the measured pressure and volume data during the separation of gas and liquid at the primary trap and tank conditions. The data reported here show that 87.92 volume units of liquid saturated at 2520 psia and 185°F will shrink to 70.02 volume units of liquid in the primary trap and then to 63.60 liquid volume units in the tank. Part B shows the method of calculating the quantities and gravities of the surface gases and tank oil.

If the primary surface trap is operating at relatively low pressures, or if the well production is going directly to the stock tank, it often is impossible to simulate the field conditions by expanding the volume of the laboratory system to the point where the cell pressure is equal to the field primary trap pressure. An alternate method of determining the gas-oil ratio and formation volume data under these conditions is illustrated in the upper right corner of Figure 48. The system in the cell first is brought into a single phase* and a portion of it forced into the small glass trap. Pressure corresponding to the field primary trap pressure is maintained by the pressure regulator. After a desired amount of liquid has accumulated in the trap, the valve at the top of the cell is closed and the trap brought to atmospheric pressure. The liquid remaining in the trap corresponds to tank oil, and the total gas measured by the wet test meter corresponds to the sum of the trap gas and tank vapors.

Differential Vaporization Tests. Differential vaporization tests are meant to represent the behavior of the portion of original liquid phase that is not produced to the surface during pressure depletion of a reservoir. As discussed in Chapter 2, the mobility of the gas phase (k_g/μ_g) becomes much greater than the mobility of the liquid phase (k_o/μ_o) after an appreciable gas saturation has been established within a rock pore. Under these conditions, a portion of the reservoir liquid phase can be thought of as being relatively immobile within the rock pore. Although the liquid may be immobile, it continuously contributes a mobile gas phase to the well production. This process constitutes a differential process; the composition of the system within the rock pore is changing with time because of the movement of the gas phase out of the pore.

The above word picture attempts to explain behavior of gas and liquid within an individual pore. A different picture is obtained if we consider phase behavior in numerous pores. As gas particle A, having just become separated from liquid particle A, travels toward the well bore, it comes in contact with liquid B. Liquid B is very similar to liquid A, having just released gas B at only a slightly lower pressure than the release at A. The combination of gas A and liquid B produces a system very close in composition to the original A system, but at a slightly lower pressure. The net result is a process comparable in many ways to a flash process, i.e., one of constant composition.

Actually, the vaporization process in a reservoir rock probably lies somewhere between the differential and flash vaporization processes. In many dissolved gas systems it makes little difference as to

* The two phases shown in the cell in Figure 48 are required for a discussion of the differential test method. For the present discussion, assume that the gas phase shown has been forced into solution.

TABLE 25. RESULTS OF FLASH TESTS ON DOUGLAS SATURATED OIL AT SURFACE TRAP CONDITIONS

Primary Trap	385 psig, 80°F
Tank	0 psig, 80°F
Saturated Oil Volume	87.92 cc at 2520 psia and 185°F
Barometer	761.7 at 75°F

PART A: Measured Data

	Primary Trap		Tank	
	Initial	**Final**	**Initial**	**Final**
Cell Pressure, psig	385		0	
Cell Temperature, °F	80			80
Total Volume, cc	261.55	70.05		63.63
Liquid Volume, cc	70.02		63.60	
Gas Volume, cc	191.53			
Gas Removed				
Volume of Collection Bottle, cc	6275			2413
Pressure, mm Hg	1.5	644.5	1.5	751.5
Temperature, °F	65			66
Gas Gravity				
Flask Pressure, mm Hg	1.0	613.5	1.0	719.0
Weight, gm	17.4158	17.5495	17.4158	17.6598
Temperature, °F	76			76

PART B: Calculated Data

Primary Gas

$$\text{Gas Volume} = 6275 \cdot \frac{520}{525} \cdot \frac{664.5 - 1.5}{760.0} = 5420 \text{ cc at } 60°\text{F and } 760 \text{ mm}$$

$$\text{Gas Gravity, } \left(\frac{\Delta m}{\Delta P}\right)_g = \frac{17.5495 - 17.4158}{613.5 - 1.0} = \frac{0.1337}{612.5} = 0.0002182$$

$$\left(\frac{\Delta m}{\Delta P}\right)_{air} \text{ at } 76°\text{F} = 0.0003157$$

$$\text{Gas Gravity} = \frac{0.0002182}{0.0003157} = 0.691$$

Tank Vapors

$$\text{Gas Volume} = 2413 \cdot \frac{520}{526} \cdot \frac{751.5 - 1.5}{760} = 2370 \text{ cc at } 60°\text{F and } 760 \text{ mm}$$

$$\text{Gas Gravity, } \left(\frac{\Delta m}{\Delta P}\right)_g = \frac{17.6598 - 17.4158}{719.0 - 1.0} = 0.000340$$

$$\left(\frac{\Delta m}{\Delta P}\right)_{air} \text{ at } 76°\text{F} = 0.0003157$$

$$\text{Gas Gravity} = \frac{0.0003400}{0.0003157} = 1.077$$

Tank Oil

Weight Pycnometer + Oil = 16.2169 gm

Weight Pycnometer = 7.9923 gm

Oil = 8.2246 gm

Temperature = 70°F
No. 5 Pycnometer Volume = 9.811 cc

$$\text{Density} = \frac{8.2246}{9.811} = 0.8383 \text{ gm per cc at } 70°\text{F}$$

$$= 0.8420 \text{ gm per cc at } 60°\text{F}$$

Specific Gravity 60/60 = 0.8428 = 36.35°API

Tank Oil Volume = 63.60 · 0.9910 = 63.03 cc at 60°F

Gas-Oil Ratio

$$\text{Primary Gas to Tank Oil Ratio} = \frac{5420}{63.03} \cdot 5.615 = 483 \text{ cu ft per bbl}$$

$$\text{Tank Vapor to Tank Oil Ratio} = \frac{2370}{63.03} \cdot 5.615 = 211 \text{ cu ft per bbl}$$

PART C: Summary

Solution gas-oil ratio = 483 + 211 = 694 cu ft per bbl tank oil

$$\text{Average solution gas gravity} = \frac{(483 \cdot 0.691) + (211 \cdot 1.077)}{483 + 211} = 0.810$$

Tank oil gravity = 36.35°API

$$\text{Bubble point formation volume at 2520 psia and 185°F} = \frac{87.92}{63.03} = 1.395 \text{ bbl per bbl tank oil}$$

which process is considered to predominate, as both processes give comparable P-V behavior over a wide range of pressure.

It is not feasible for laboratory tests to correspond to true differential vaporization processes. To do so requires that each infinitesimal gas bubble be removed from the cell as it is formed. Rather, in the laboratory, a series of stepwise flash vaporization conditions is imposed upon the original system and the gas formed at each step is removed as a unit. This stepwise process is called a differential process in practically all PVT work concerned with dissolved gas systems.

Figure 51 illustrates the process of conducting a differential vaporization test at 185°F. In this particular test, gas was removed from the system in nine steps, the first step occurring at 2203 psia and the last step occurring at 14.7 psia. Ordinarily, no effort is made to remove the gases at particular pressures, although the pressure increment between steps normally is limited to 300 psi.

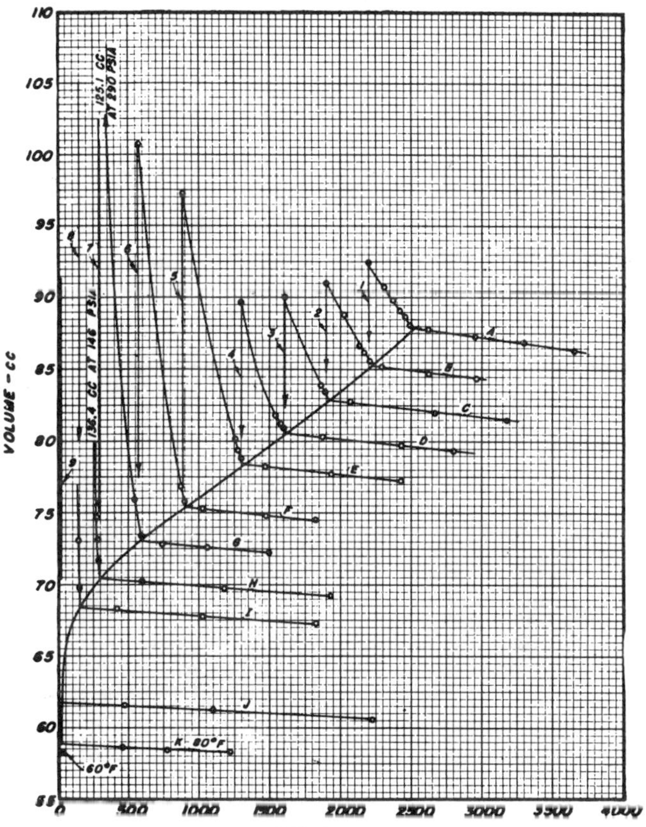

Figure 51—Pressure-Volume Relations of Douglas Saturated Oil. Differential Vaporization at 185°F

A portion of the P-V behavior of the original system was determined prior to removal of the first gas increment at 2203 psia. This behavior is shown by curve A. After bringing the original system to equilibrium at 2203 psia and measuring the total volume of the system (92.48 cc), the gas phase was removed from the top of the cell and its quantity measured by collecting it in previously evacuated bottles. The cell pressure was maintained as close as possible to 2203 psia during the gas removal process by injecting mercury into the bottom of the cell. In Figure 51, the process of removing the gas from the cell is designated by the vertical arrow connecting system A to system B.

After removal of the first gas increment, a portion of the P-V curve of system B was determined. System B was found to have a bubble point pressure of 2230 psia and a bubble point volume of 85.25 cc. The fact that the bubble point pressure of system B was greater than 2203 psia indicates that a small amount of gas was left in the cell at the completion of the previous gas withdrawal step. Ordinarily, all the gas phase is removed from the cell at each step, so that the bubble point pressure of the new system agrees with the pressure at which the gas withdrawal process is carried out.

The data in Table 26 show that 1017 cc of gas, measured under standard conditions, was removed in changing from system A to system B. The 6.75 cc volume that the 1017 cc of gas occupied in the cell at 2203 psia and 185°F is the differential between the bubble point volume of system B (85.25 cc) and the volume of system A at 2230 psia (92.00 cc). The compressibility factor of the gas which came out of solution between 2520 and 2230 psia is calculated at 2230 psia as follows:

$$\frac{P_1V_1}{Z_1T_1} = NR = \frac{P_2V_2}{Z_2T_2} \qquad (42)$$

where subscripts 1 and 2 indicate standard conditions and reservoir conditions, respectively.

Solving Equation (42) for Z_2 gives

$$Z_2 = \frac{P_2V_2T_1Z_1}{P_1V_1T_2} \qquad (43)$$

As Z_1 is very close to 0.998 at 14.7 psia and 60°F,

$$Z_2 = \frac{2230 \cdot 6.75 \cdot 520 \cdot 0.998}{14.7 \cdot 1017 \cdot 645} = 0.810$$

The gravity of the gas which came out of solution between 2520 and 2230 psia is shown in column 6 of Table 26. These gravities were determined by the method outlined in the previous section.

TABLE 26. RESULTS OF DIFFERENTIAL VAPORIZATION TESTS AT 185°F
Initial Saturated Oil Volume = 87.92 cc at 185°F

1	2	3	4	5	6	7
		Volume of Gas Removed				
Step	Pressure PSIA	cc at 14.7 PSIA and 60°F	cc in Cell	Liquid Phase Volume cc	Gas Gravity Air=1	Gas Phase Compressibility Factor Z
0	2520	0	0	87.92		
1	2230	1017	6.75	85.25	0.743	0.810
2	1930	964	7.50	82.87	0.740	0.823
3	1617	940	9.21	80.54	0.742	0.868
4	1310	984	10.86	78.34	0.752	0.872
5	907	1172	20.21	75.34	0.766	0.858
6	588	922	25.77	72.93	0.822	0.899
7	300	903	52.5	70.40	0.938	0.955
8	147	596	....	68.46	1.159	.....
9	14.7	1545	....	61.89	1.73	.....
14.7 PSIA & 60°F				58.26		
	TOTAL	8953				

Residual Oil Specific Gravity — 0.8528 60/60 — 34.4° API

The shrinkage that the liquid phase undergoes because of gas removal is indicated by the curve connecting the several bubble points shown in Figure 51. System J was the final liquid remaining in the cell at 14.7 psia and 185°F. System K is the same material as system J, but measured at a temperature of 80°F. The volume of the residual liquid left in the cell at 14.7 psia and 60°F was obtained by correcting the 80°F volume for the effect of temperature. A residual oil volume of 58.26 cc was obtained by this means.

Table 26 presents a tabulation of results of the differential test at 185°F. The cumulative weight of hydrocarbons in the cell at each pressure can be calculated from data presented in this table. The weights are as shown in column 5 of Table 27. These weight data in column 5 with the volumetric data pertaining to the bubble points of each system give the density of the bubble point liquid at each pressure. The accuracy of the laboratory tests can be checked by comparing the density of the original system at 2520 psia and 185°F as calculated from differential test data and flash test data. The value 0.6854 gm per cc shown in Table 27 is less than one per cent different from the 0.6910 gm per cc calculated from flash data of Table 25. This is considered a good check.

The gravity of the gas which came out of solution at each step is listed in Table 26 and shown graphically by curve A of Figure 52. As the measured gravities represent an average of the gases coming from solution over a pressure interval, they are plotted as horizontal line segments. For example, in Table 26, the measured gas gravity of 0.743 of the 1017 cc removed during step one is the average

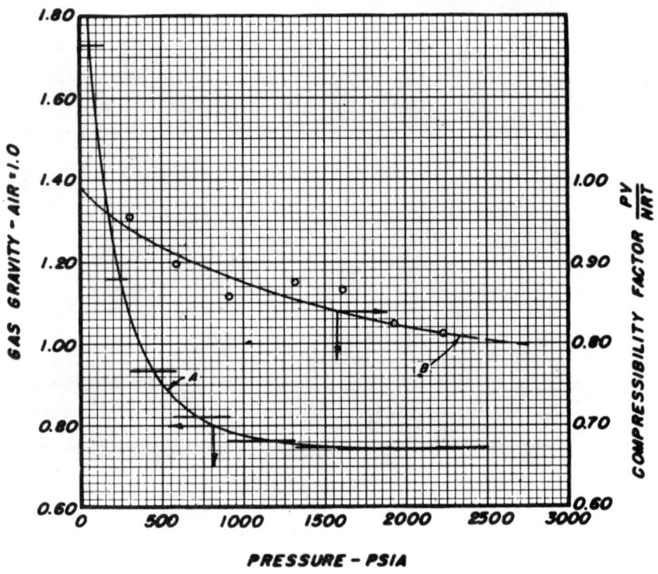

Figure 52—Gravity and Compressibility of Liberated Gas. Differential Liberation at 185°F

of the gases coming out of solution between 2520 and 2230 psia. The instantaneous gas gravity, curve A of Figure 52, is obtained by drawing the best curve through the horizontal line segments.

Table 27. Calculation of Liquid Phase Specific Volume from Differential Test Data

Temperature 185°F

1	2	3	4	5	6	7	8
Bubble Point Pressure PSIA	Quantity of Gas in Solution		Weight of Residual Oil in Systems	Bubble Point Weight and Volume of Systems		Bubble Point Density of Systems	
	cc at 14.7 PSIA and 60°F	gram	gram	gram	cc	gram per cc	cu ft per lb
14.7*	0	0	49.632	49.632	58.26	0.8519	0.01883
14.7	0	0	↓	49.632	61.89	0.8019	0.02000
147	1545	3.266		52.898	68.46	0.7727	0.02075
300	2141	4.110		53.742	70.40	0.7633	0.02102
588	3044	5.143		54.775	72.93	0.7510	0.02135
907	3966	6.068		55.700	75.34	0.7393	0.02168
1310	5138	7.164		56.796	78.34	0.7249	0.02210
1617	6032	7.985		57.617	80.54	0.7153	0.02240
1930	6972	8.836		58.468	82.87	0.7055	0.02272
2230	7936	9.706		59.338	85.25	0.6960	0.02302
2520	8953	10.629		60.261	87.92	0.6854	0.02338

* At 60°F

Notes on Calculation Method
Column 3. Weight of gas = 0.001222 · cc · gravity

Each compressibility factor listed in Table 26 is that of a different composition gas, although above 1500 psia the gas gravity data indicate only minor differences in composition of the gases. As the volumes of the gas phases within the cell were measured at specific pressures, the compressibility factors must be plotted at the pressure at which the measurement was made. For example, the average of the gases coming out of solution between 2520 and 2230 psia has a calculated compressibility factor of 0.810 at 2230 psia and 185°F. This fact should be kept in mind when comparing a smooth curve such as curve B in Figure 52 with compressibility factor curves of constant composition gases.

Flash Vaporization of Dissolved Gas Systems at Surface Conditions

It is easy to see from the data in Tables 25 and 26 that a barrel of reservoir oil undergoing differential vaporization at reservoir temperatures near 200°F will give different quantities of gas and oil than will the same barrel undergoing flash vaporizations at the lower pressures and temperatures existing at surface traps. As an example, the data in Table 26 show that, under the reservoir conditions of 185°F, 87.92 cc of 2520 psia bubble point liquid produced a total of 8953 cc of gas and 58.26 cc of residual liquid on undergoing differential vaporization at 185°F from 2520 psia to atmospheric pressure. Table 25 shows that a comparable 87.92 cc of bubble point liquid produced only 7790 cc of gas and 63.03 cc of liquid when separated by a two-stage surface trapping arrangement. Thus, the gas-oil ratio of the 2520 psia bubble point liquid can be expressed as $\frac{8953}{58.26} \cdot 5.615 = 864$ cu ft per bbl or $\frac{7790}{63.03} \cdot 5.615 = 694$ cu ft per bbl, depending on which separation is being considered. Obviously, the exact processes by which the gas and oil separations are made must be known when speaking of gas-oil ratio and liquid volume shrinkage. This can be done by using the term "residual oil" to indicate the liquid at 14.7 psia and 60°F that results from a differential process carried out at, or near, the reservoir temperature. "Tank oil" is the term for the liquid at 14.7 psia and 60°F that results from a flash process at, or near, surface temperatures. The terms "saturated oil" and "bubble point liquid" are common to both processes and serve as the common basis for conversions between differential and flash processes. The ratio of the volume of a system to the volume of residual oil, or the volume of saturated oil, is called the "relative volume" of the system. The ratio of system volume to tank oil volume is called the "formation volume." Thus, in the example outlined above, the relative volume of the bubble point liquid at

TABLE 28. EFFECT OF SURFACE TRAP PRESSURE ON QUANTITY AND PROPERTIES OF SURFACE PRODUCTION

1	2	3	4	5	6	7	8	9
Trap Pressure PSIG	Surface Gas Released cu ft at 14.7 PSIA and 60°F			Surface Gas Gravities Air=1.0		Tank Oil Produced at 14.7 PSIA and 60°F		Total Gas-Oil Ratio cu ft per bbl tank oil
	Trap	Tank	Total	Trap	Tank	bbl	°API	
385	346	151	497	0.691	1.077	0.717	36.4	694
285	387	109	496	0.709	1.142	0.721	36.6	688
185	416	87	503	0.732	1.24	0.721	36.4	697
0	558	0	558	0.977	.....	0.680	34.4	820

Basis: One barrel of oil saturated at 2520 PSIA and 185° F
Trap and tank operating at 80° F

TABLE 29. EFFECT OF SURFACE TRAP TEMPERATURE ON QUANTITY AND PROPERTIES OF SURFACE PRODUCTION

1	2	3	4	5	6
Separation Temperature °F	Total Gas-Oil Ratio cu ft per bbl tank oil	Formation Volume of Saturated Reservoir Oil bbl per bbl tank oil	Tank Oil Gravity °API	Surface Gas Richness Gal C_4 + per MCF	
				Trap	Tank
Primary Trap at 385 PSIG					
100	710	1.425	35.6	0.99	8.1
80	694	1.395	36.4	0.71	5.9
60	679	1.369	37.2	0.45	3.8
Primary Trap at 185 PSIG					
100	713	1.415	35.6	1.40	9.2
80	697	1.387	36.4	1.08	6.9
60	682	1.362	37.2	0.78	4.6

Basis: One barrel of oil saturated at 2520 PSIA and 185° F
Trap and Tank operating at indicated pressures and temperatures

2520 psia and 185°F is $\frac{87.92}{58.26} = 1.509$ bbl per bbl residual oil. The formation volume of the bubble point liquid is $\frac{87.92}{63.03} = 1.395$ bbl per bbl tank oil.

While it is apparent that the two different methods of separating gas and oil, mentioned above, can give different gas-oil ratios and liquid phase shrinkages, it is not so obvious that relatively small changes in surface trap pressures and temperatures also can cause comparable differences. Table 28 shows the result of flashing one barrel of reservoir oil saturated at 2520 psia and 185°F through a primary trap-tank setting operating at 80°F. By varying the primary trap pressure between 385 and 0 psig, the gas-oil ratio of the same material varies between 688 and 820 cu ft per bbl. Likewise, the formation volume of the original barrel of saturated reservoir oil varies from 1.387 to 1.47, depending on the surface pressures at which the barrel of reservoir oil is separated.

The temperature at which the surface separation is accomplished also affects the amount of surface gases and tank oil obtained from a unit volume of reservoir oil. The data in Table 29 show the effect of varying separation temperature between 60° and 100°F. The temperature effect is greatest for high gravity production but essentially disappears for tank oils less than 20° API.

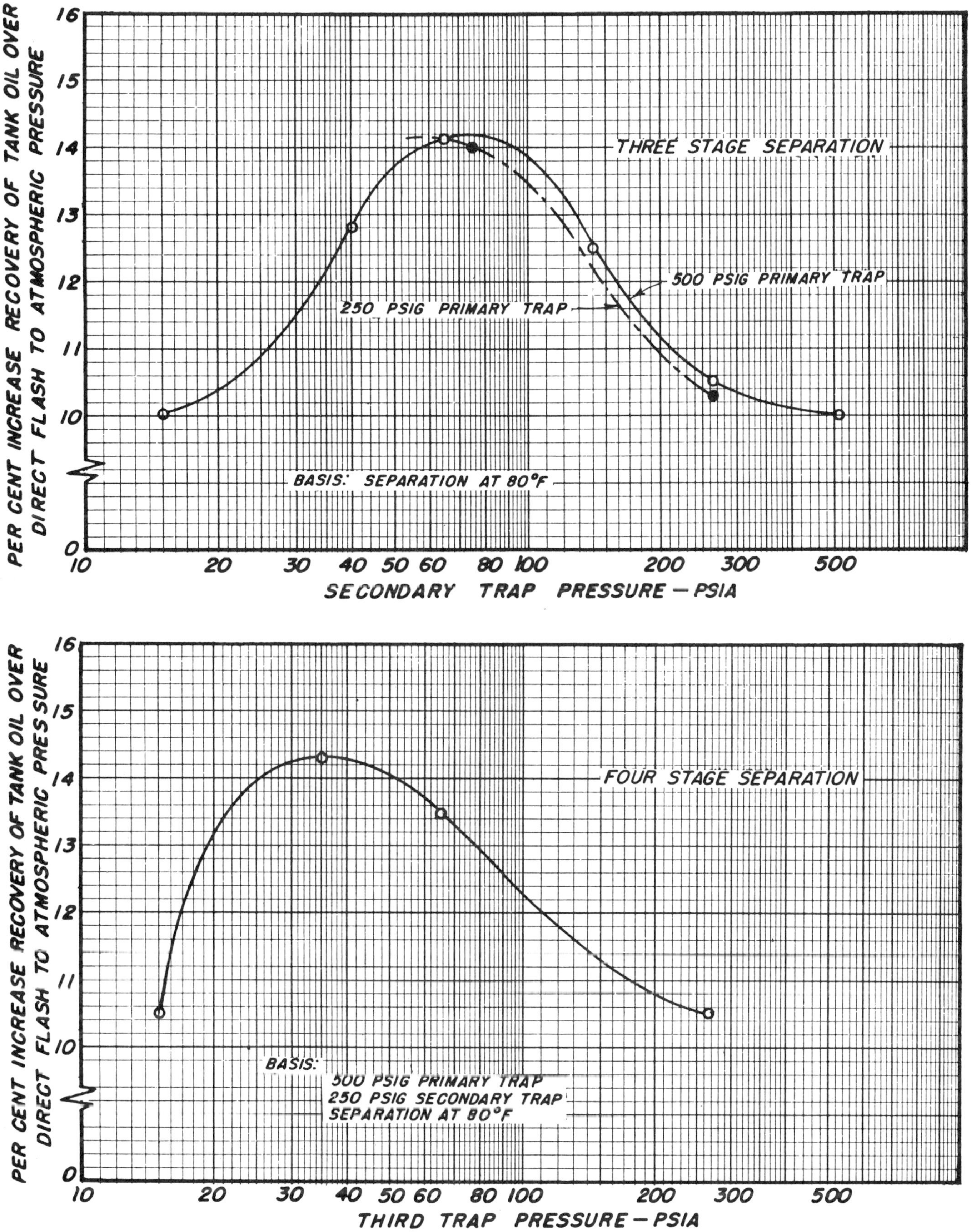

Figure 53—Effect of Stage Separation on Tank Oil Recovery

A third factor which can affect the quantity of surface products is the number of traps used in series to separate the gas and oil. Increasing the number of stages will result in more tank oil of higher gravity and less surface gases. For any particular well effluent and temperature, there are optimum pressures at which the traps can be operated in order to yield the maximum tank oil or, if desired, the maximum amount of gas. An example of the effect of stage separation on tank oil recovery is shown in Figure 53.* The upper half of Figure 53 shows that with a primary trap operating at 515 psia, the optimum secondary trap pressure would be 75 psia. Operating with this combination of trap pressures would result in recovering 14.2 per cent more tank oil than would be recovered if the well production were flashed directly to the tank. The chart also shows that essentially the same recovery would be obtained if the primary trap pressure were reduced to 265 psia and the secondary trap operated at 55 psia. The lower part of Figure 53 shows the effect of maintaining a 500 psig primary and 250 psig secondary trap and varying the pressure of the third-stage trap. It can be seen that for the particular fluid under discussion the addition of a third trap yields little additional oil in the tank over that accomplished by operating with a three-stage separation.

Data presented in Figure 53 express in general the behavior of naturally occurring dissolved gas systems when processed through surface traps. The greatest incremental return in tank oil is obtained in going from a single-stage to a two-stage separation. A somewhat smaller incremental return results from the addition of a third stage. Field experience has shown that a fourth stage seldom is justified on the basis of tank oil recovery. Operational considerations, however, may warrant four-stage separation. The trap pressures yielding the optimum amount of tank oil are dependent upon the composition of the system being separated and are determined best by laboratory experimentation or by calculation. In a two-stage separation of dissolved gas systems, the maximum quantity of tank oil usually is obtained when operating the first-stage trap in the pressure range 150 to 250 psig. Three-stage separation often calls for increasing the first stage pressure to 400 or 500 psig with the second stage operating in the region of 60 psig.

A precise study of the pressure-production behavior of a reservoir requires the gas solubility and formation volume data used in the study to be evaluated under conditions representative of the field operating conditions. This usually requires specific laboratory PVT tests to determine the appropriate properties of the reservoir fluids. However, in many instances, inaccuracies of several per cent are allowable in the values of bubble point pressure, gas solubility, or formation volumes and, as a result, use can be made of general correlations of these properties. A correlation of the formation volume of bubble point liquids already has been given in Figure 27, and a correlation of the total gas plus liquid phase formation volume is shown in Figure 39. In Figure 54[7] is shown a correlation of the bubble point pressure of a dissolved gas system with dissolved gas-oil ratio, dissolved gas gravity, temperature, and tank oil gravity. A calculating chart based on this correlation is given in the Appendix.

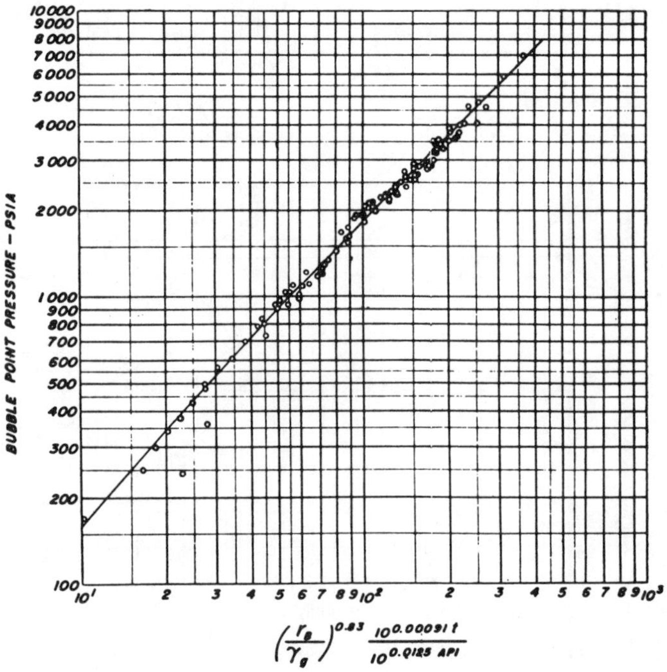

Figure 54—Correlation of Bubble Point Pressure with Dissolved Gas-Oil Ratio, Dissolved Gas Gravity, Tank Oil Gravity, and Temperature

* Not Douglas production but a dissolved gas system having properties close to those of the Douglas system.

The correlation shown in Figure 54 only indirectly takes into account the effect of separation pressure and temperature on gas-oil ratio. The properties of the systems comprising the correlation were determined, in general, under conditions of two-stage flash vaporization at 100°F, with the first-stage pressure in the range from 150 to 400 psia. This temperature and pressure range reflects the general field trap operating conditions in California and, fortunately, falls in the range wherein the effect of trap pressure on gas-oil ratio is small. (See column 9, Table 28.) However, the correlation often works well even outside the trap pressure range mentioned above. For example, the bubble point pressure of the Douglas saturated oil was calculated correctly to be 2520 psia at 185°F by using the separation data shown in Table 28 for the 385 psig primary trap condition. Using the separation data shown for a 0 psig primary trap gave a calculated bubble point pressure of 2625 psia.

Similarly, the effect of trap separation conditions on the formation volume of bubble point liquids is taken into account indirectly in Figure 27 by the dissolved gas-oil ratio and gas gravity. For example, the 385 psig data shown in Table 28 give a calculated formation volume of 1.41 from Figure 27. The true value is 1.395. The 0 psig data give a calculated value of 1.50 as compared to a true value of 1.47. This agreement is considered reasonable in view of the 18 per cent difference in gas-oil ratio.

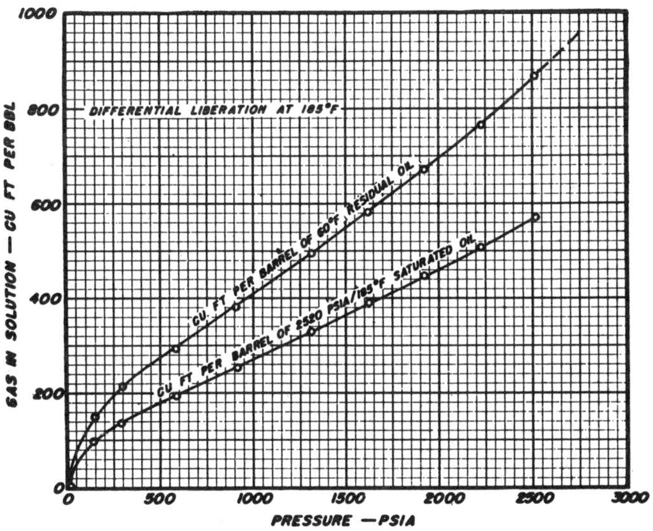

Figure 55—Gas Solubility vs Pressure
Douglas Saturated Oil
Differential Liberation at 185°F

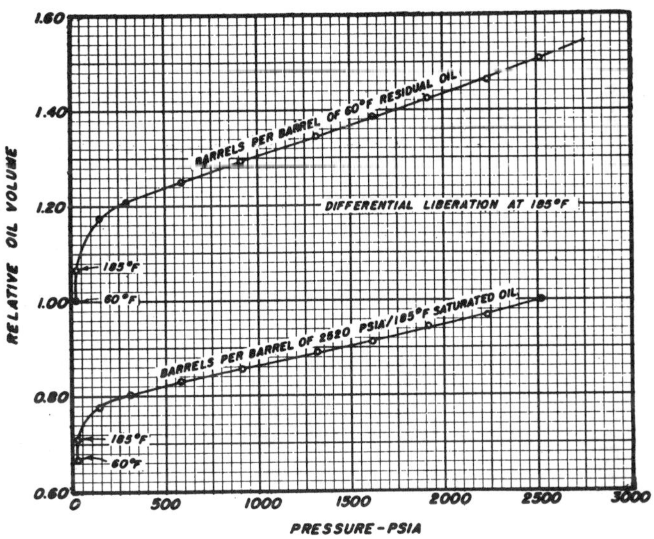

Figure 56—Relative Oil Volumes vs Pressure
Douglas Saturated Oil
Differential Liberation at 185°F

Differential Vaporization of Dissolved Gas Systems at Reservoir Conditions

Gas solubility determined by differential test methods is expressed most often in terms of cubic feet per barrel of residual oil or of initially saturated oil. Relative oil volumes also are commonly based on a unit volume of residual oil or of saturated oil. Data on the solubility and relative oil volumes of the Douglas saturated oil at 185°F are shown in Figures 55 and 56. The values shown in these figures are calculated from the data given in columns 2 and 6 of Table 27. It will be appreciated that in each figure the values of the two curves at any pressure are in the ratio of 58.26 to 87.92, which are the respective volumes of the residual oil at 60°F and the 2520 psia saturated oil at 185°F.

It often is necessary to compare the results of differential tests on similar samples tested by different laboratories. It is quite possible that the comparison will be complicated by a difference in the bubble point pressures of the two samples. When such a difference does occur, it is obviously incorrect to compare the gas solubilities and relative volumes on the basis of unit volumes of saturated oils. Rather, one should make comparisons on the basis of a unit of residual oil or some saturated oil that is common to both systems. When this is done, one usually will find that the solubility and relative volume of the two samples will be in agreement.

At times the solubility and relative volume curves of two similar systems will not be in agreement when based on a barrel of residual oil, but both sets of curves will be displaced from each other by a constant percentage. When this occurs, it is usually because of an error in the measurement of one of the residual oil volumes. If there is good agreement between the relative oil volume curves but poor agreement in the solubility curves, the possible errors are indicated to be in the measurement of the liberated gas quantities.

Fortunately, differential gas solubility curves and relative oil volume curves are not sensitive to the pressure at which the differential gas withdrawal process was initiated. As pointed out in Chapter 2, bottom-hole liquid samples or recombined surface samples often are quite different in composition from the liquid phase within the reservoir rock. Nevertheless, good estimates of the properties of the reservoir liquid can be made from extrapolated properties of the laboratory samples. As a matter of fact, the extrapolation may not involve going just from well bore conditions to average reservoir conditions, but often involves extrapolating from the present day well fluid properties to original reservoir conditions. This will be illustrated in the following section wherein the properties of the original Douglas reservoir oil at 2750 psia and 185°F are calculated from the measured properties of the 2520 psia bubble point liquid obtained in a bottom hole sampler.

Adjusting Laboratory Data to Field Conditions

The method of applying the properties of dissolved gas systems to reservoir behavior calculations will be discussed in the next chapter. For the moment, it is sufficient to know that one of the methods to be discussed requires a P-V curve at reservoir temperature of the original liquid phase in the reservoir, the solubility of gas in the original reservoir liquid when separated under the field trapping conditions, and the formation volume factor of the original reservoir liquid. The second material balance method requires the flash gas solubility-pressure and formation volume factor-pressure relations of the reservoir liquid phase, both determined under surface separating conditions. All these factors can be obtained by modifying the data presented thus far.

Pressure-Total Volume Relations. An experimental method of developing a system having a saturation or bubble point pressure of 2750 psia at 185°F is to add sufficient gas to the bottom hole sample to raise its bubble point pressure from 2520 psia to the desired value. The added gas should have, of course, a composition as close as possible to the gas that originally came out of solution in the reservoir liquid between 2750 and 2520. One source of such a gas is casing gas taken at the time the bottom hole sample is taken. A second choice of gas would be from the primary trap. However, in many instances, the composition of the primary trap gas is different from the composition of the free reservoir gas and consequently has limited applicability.

The mathematical approach to the problem is to use the data shown in Figures 52, 55, and 56 to obtain rough values for the new P-V curve having the desired bubble point conditions, and then to smooth the rough values by use of the previously mentioned "*Y*" correlation. The steps are as follows:

(1) On the basis of a unit barrel of residual oil, the liquid phase relative volume curve of system A is drawn as in Figure 57. The curve is extrapolated to 2750 psia, giving a value of 1.548 bbl per bbl of residual oil at this pressure. The extrapolation also is shown in Figure 56.

(2) The P-V curve of system A (2520 psia bubble point) is constructed in terms of the unit barrel of residual oil. This is done by dividing the total volumes of system A, listed in Table 24, by the volume of the residual oil, 58.26 cc. For example, at 1710 psia, the relative volume of system A is $\frac{105.02}{58.26} = 1.803$. This is shown in Figure 57.

(3) Extrapolation of the upper solubility curve in Figure 55 from 2520 to 2750 psia indicates that system B will contain an additional 92 cu ft of gas per bbl of residual oil. (995 −

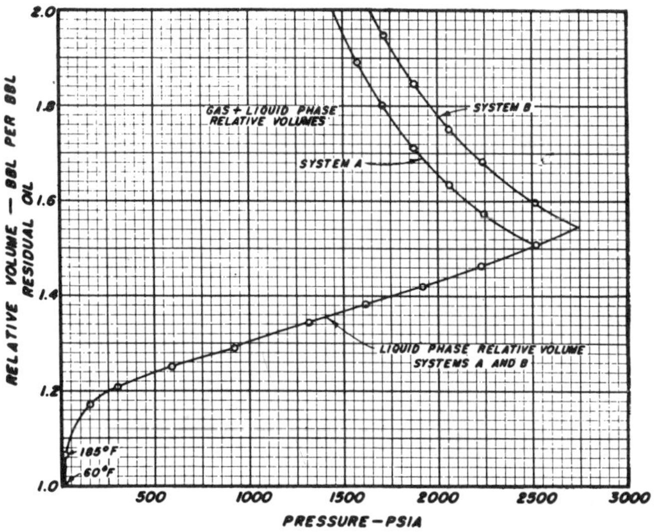

Figure 57—Construction of P-V Curve for System B from Data of System A

863 = 92). These additional 92 cu ft of gas, when compressed to pressure P and 185°F, will occupy

$$\frac{92}{5.615} \cdot \frac{14.7}{P} \cdot \frac{645}{520} \cdot Z = 298.5 \frac{Z}{P} \text{ bbls of space.}$$

This incremental volume of gas added to system A's relative volume curve yields the desired P-V curve for system B. The P-V curve of system B then is smoothed by a "Y" correlation. These calculations are shown in detail in Table 30.

The above procedure is based on the assumptions that the relative liquid volume curves of systems A and B are the same and that the compressibility factor of the 92 incremental cu ft of gas can be represented by the curve shown in Figure 52. The first assumption generally is true when there is not too large a difference in the bubble point pressures of the two systems. The second assumption is good over the range of pressure in which the composition of the gas coming from solution is close to the composition of the incremental gas. In this particular instance, the gravity of the gas (which reflects the composition of the gas) remains relatively constant down to about 1000 psia.

A second method of calculating a P-V curve from differential vaporization data is illustrated in Table 31. The volume of free gas present at any pressure is calculated from the solubility curve and a compressibility factor determined from the weighted gas gravity. The free gas volume when added to the liquid phase relative volumes gives the desired two-phase P-V curve. The P-V curve then can be smoothed by the usual "Y" correlation. This method is not as accurate as the previous one as it depends in part on a gas phase compressibility factor determined from reduced pressure and temperature correlations. The example shown in Table 31 has been limited purposely to calculating relative volumes of system A in order that a comparison

Table 30. Method of Calculating P-V Curve for System B from Data on System A

1	2	3	4	5	6	7	8
Pressure PSIA	Relative Volume System A bbl per bbl residual oil	Free Gas Compressibility Factor	Free Gas Volume $=298.5\frac{Z}{P}$ bbl	Relative Volume System B bbl per bbl residual oil	"Y"	Smoothed "Y"	Smoothed Relative Volume System B bbl per bbl residual oil
2750	.....	.....	......	1.5480	.....	.....	1.5480
2520	1.5091	0.804	0.0952	1.6043	2.5102	2.7365	1.5996
2247	1.5751	0.812	0.1079	1.6830	2.5671	2.6100	1.6808
2061	1.6345	0.818	0.1185	1.7530	2.5245	2.5245	1.7530
1876	1.7123	0.826	0.1315	1.8438	2.4381	2.4381	1.8438
1710	1.8026	0.833	0.1455	1.9481	2.3532	2.3532	1.9481
1572	1.8942	0.840	0.1595	2.0537	2.2939	2.2939	2.0537
1361	2.0786	0.852	0.1870	2.2656	2.2016	2.2016	2.2656
1148	2.3292	0.865	0.2247	2.5539	2.1476	2.1040	2.5747
926	2.8027	0.880	0.2838	3.0863	1.9822	2.0025	3.0999

Notes on Calculation Method

Column 2. Relative volume based on barrel of 60° F residual oil. Data from Table 24.

Column 3. Compressibility factor data from Figure 52.

Column 5. Value at 2750 PSIA obtained by extrapolation of curve shown in Figure 57.

TABLE 31. METHOD OF CALCULATING P-V CURVE FROM DIFFERENTIAL VAPORIZATION DATA

1	2	3	4	5	6	7	8
Pressure PSIA	Free Gas cu ft per bbl residual oil (864.0−S)	Weighted Average Free Gas Gravity	Free Gas Compressibility Factor Z	Free Gas Volume $=3.248\,\frac{(864.0-S)Z}{P}$ bbl	Liquid Phase Relative Volume	Total Gas Plus Liquid Relative Volumes bbl per bbl residual oil: Calculated	Experimental
2520	0	.....	.....	.....	1.509	1.509	1.509
2230	98.1	0.743	0.816	0.117	1.463	1.580	1.579
1930	191.3	0.742	0.826	0.266	1.423	1.689	1.688
1617	281.8	0.742	0.840	0.476	1.383	1.859	1.858
1310	368.0	0.744	0.857	0.782	1.344	2.126	2.135
907	480.5	0.749	0.895	1.540	1.293	2.833	2.850
588	570.0	0.760	0.927	2.920	1.251	4.171	4.245
300	657.0	0.783	0.958	6.815	1.207	8.022	8.310
147	715.0	0.814	0.980	15.50	1.174	16.67	17.22
15	864.0	0.973	.....	.....	1.062	.....	.....

Basis: One barrel 60° F residual oil

Notes on Calculation Method

Column 2. Free gas from solubility data of column 2, Table 27, expressed as cu ft per bbl of residual oil, from Figure 55
Column 3. Data from columns 3 and 6, Table 26
Column 4. Via pP_C and pT_C Figure 17 and Z from Figure 15
Column 5. Volume of free gas at P and T $=\frac{(864.0-S)}{5.615}\cdot\frac{14.7}{P}\cdot\frac{645}{520}\cdot Z = 3.248\,\frac{(864.0-S)Z}{P}$
Column 6. From Figure 56
Column 7. Sum of columns 5 and 6
Column 8. From Y curve, Figure 50, converted into terms of a barrel of 60° F residual oil

can be made of calculated and experimental values. This comparison is shown in columns 7 and 8.

The third method of calculating the P-V curve of a system is by use of the generalized correlations given in Figures 39 and 54, or by use of the respective calculating chart in the Appendix. In the example given in Table 32, the gas plus liquid phase formation volume data in column 2 were obtained from the two-phase formation volume calculating chart. The gas-oil ratio of the particular system was evaluated by making use of the calculating chart and a bubble point pressure of 2750 psia at 185°F. The dissolved gas gravity was assumed to be 0.81, and the tank oil gravity was taken as 36.4° API. Use of the calculating chart with these properties of the surface gas and oil gave a solution gas-oil ratio of 770 cu ft per bbl of tank oil. Column 3 in Table 32 refers the formation volume data to the reference condition of 2750 psia. The data of columns 1 and 3 give the calculated values of "Y" shown in column 4. The smoothed "Y" data in column 5 were obtained by constructing a straight line through the data of column 4, less weight being given to the calculated values of "Y" near the bubble point of the system than at pressures between 2200 and 1300 psia. Column 6 presents smoothed values of the two-phase volumes relative to the bubble point volume. The data of column 6 have been referred to a unit of residual oil in column 7 to facilitate comparison with the relative volume data of system B given in Table 30. This third method of calculating a P-V curve of a particular system has considerable merit, as it can be used when direct laboratory data are not available.

Pressure-Solubility Relations. When using the Schilthuis[41] form of material balance, it is necessary to know the solution gas-oil ratio of only the original liquid phase in the reservoir. Consequently, in many routine PVT tests a solution gas-oil ratio (flash) of only one system ordinarily is determined. In many instances, the gas-oil ratio reported is that of the bottom-hole sample. It remains for the engineer to convert this ratio to a value representative of the original reservoir liquid phase. More recently, Tarner[42] suggested a form of material balance that requires the solution gas-oil ratio of the reservoir liquid phase at all pressures below the original reservoir pressure. Thus, to use Tarner's method of material balance, it is necessary to be able to convert solubilities determined by differential methods to flash solubilities.

It is not too difficult to determine experimentally

Table 32. Method of Calculating P-V Curve from Generalized Two-phase Formation Volume Correlation[7]

1	2	3	4	5	6	7
Pressure PSIA	Formation Volume bbl per bbl tank oil	Volume Relative to Volume at 2750 PSIA	"Y"	Smoothed "Y"	Smoothed Volume Relative to Volume at 2750 PSIA	Smoothed Relative Volume bbl per bbl residual oil
2750	1.47	1.000	——	——	1.0000	1.5480
2520	1.51	1.027	3.380	2.865	1.0318	1.5972
2247	1.58	1.075	2.985	2.718	1.0824	1.6756
2061	1.66	1.129	2.592	2.613	1.1279	1.7460
1876	1.75	1.191	2.439	2.510	1.1856	1.8353
1710	1.84	1.252	2.413	2.420	1.2513	1.9370
1572	1.93	1.313	2.394	2.343	1.3198	2.0431
1361	2.14	1.456	2.238	2.230	1.4576	2.2564
1148	2.43	1.653	2.137	2.110	1.6613	2.5717
926	2.83	1.925	2.129	1.990	1.9898	3.0802

Notes on Calculation Method

Column 2. Formation volume of gas plus liquid phase from generalized property chart in Appendix
Column 3. Values of column 2 divided by 1.47
Column 4. Calculated from values in column 3 and Equation 37
Column 5. Smoothed values from straight line relation of Y vs P
Column 6. From Equation 37 and smoothed "Y" values in column 5
Column 7. Based on one barrel of residual oil or 1.5480 barrel of 2750 PSIA saturated oil. Column 7 = column 6 · 1.5480

the flash pressure-solubility relations of liquids corresponding to the reservoir liquid. For example, in addition to determining the gas-oil ratio of 2520 psia saturated liquid, as shown in Table 28, it is feasible to use a second portion of bottom-hole sample and progress from 2520 psia to, say, 1930 psia by a differential path at 185°F, and then separate the 1930 psia saturated oil at 80° F by stages at 385 and 0 psig. This procedure could then be repeated a third and a fourth time to determine the gas in solution at, say, 1310 and 588 psia, when separated under the prescribed surface trap conditions. Each time it would be necessary to start with 2520 psia saturated oil and progress to the lower pressure by the same differential path.

A second experimental method is to start with enough 2520 psia saturated oil to handle both the differential and flash tests. The method is to proceed with the differential tests at reservoir temperature interspersed with flash tests at trap conditions on small portions of the main sample. This process requires a very large sample and complicated auxiliary equipment for the flash tests, and generally is not carried out for these reasons.

A satisfactory method of calculating a flash solubility curve is by use of the correlation shown in Figure 54. Over the pressure range from 1000 to 8000 psia, the equation of the curve is closely approximated by

$$P_B = 18\left(\frac{r_B}{\gamma_g}\right)^{0.83} \cdot \frac{10^{\,0.00091t}}{10^{0.0125\,\mathrm{API}}} \tag{44}$$

Equation (44) can be used to calculate the effect of pressure on solubility if the manner is known by which the gravity of the dissolved gas and the produced tank oil gravity varies with pressure. Where the five variables used in Equation (44) are known for one system, the relation of the variables can be used to calculate the pressure-solubility relation of a similar system. For example, data given in Table 25 show that the Douglas bottom-hole sample, which had a bubble point pressure of 2520 psia at 185°F, produced 694 cu ft of 0.81 gravity surface gas per bbl of 36.35° API tank oil when separated by a particular two-stage process at 80°F. The solution gas-oil ratio of the original reservoir oil, saturated at 2750 psia and 185°F, is calculated by writing Equation (44) for each bubble point condition and dividing one equation by the other; thus

$$\frac{2750}{2520} = \left(\frac{r_B}{694}\right)^{0.83}\left(\frac{0.810}{\gamma_g}\right)^{0.83} \cdot \frac{10^{0.0125\cdot 36.35}}{10^{0.0125\cdot \mathrm{API}}} \tag{45}$$

where r_B, γ_g, °API represent the as yet unknown properties of the 2750 psia production.

A good assumption is that the gravity of the tank oil obtained from reservoir oil saturated at 2750 psia and 185°F will be the same as the gravity of tank oil from reservoir oil saturated at 2520 psia and 185°F. With this assumption, Equation (45) can be simplified further to

$$r_B = 694\left(\frac{\gamma_g}{0.810}\right)\left(\frac{2750}{2520}\right)^{\text{exp.}\frac{1}{0.83}} \tag{46}$$

or

$$r_B = 952\gamma_g \tag{47}$$

The assumption that tank oil gravity remains relatively constant for reservoir pressures above 1000 psia is reasonable, and is in agreement with field observations. (At low reservoir pressures, the rich reservoir gas, when produced simultaneously with saturated reservoir oil, often condenses in the surface traps and causes an increase in tank oil gravity.)

A second equation relating produced solution gas-oil ratio with solution gas gravity can be set up by making use of the results of differential vaporization tests. Referring to Figure 52, it is reasonable to expect that had we been able to sample the original reservoir oil, the gravity of gas coming out of solution at 185°F between 2750 and 2520 psia would have been 0.745. In terms of surface gas production, the 2750 psia saturated oil can be thought of as containing (r_B − 694) units of 0.745 gravity gas, in addition to the 694 units of 0.81 gravity gas contained in the 2520 psia saturated oil. Therefore, the average gas gravity of the 2750 psia saturated oil system can be expressed:

$$\gamma_g = \frac{694 \cdot 0.81 + (r_B - 694)0.745}{r_B} \tag{48}$$

where 694 = solution gas-oil ratio of the 2520 psia saturated oil
0.81 = total gravity of surface gases produced from 2520 psia saturated oil
r_B = solution gas-oil ratio of 2750 psia saturated oil
γ_g = total gravity of surface gases produced from 2750 psia saturated oil
0.745 = average gravity of reservoir gas that would be liberated between bubble point pressures of 2750 and 2520 psia. Value obtained by extrapolation of curve A, Figure 52.

Solving simultaneously Equations (47) and (48) gives the following quadratic equation:

$$r_B^2 - 709r_B - 42945 = 0 \tag{49}$$

The particular solution of Equation (49) is that r_B = 765 cu ft per bbl of tank oil and that the solution gas gravity is 0.804.

The solution gas-oil ratios and solution gas gravities of 2000, 1500, and 1000 psia saturated oil calculated by similar equations are 538, 395, and 250 cu ft per bbl and 0.830, 0.862, and 0.920 gravity respectively. These calculated values, together with the 2750 psia saturated oil values, are shown as the solid black squares in Figures 58 and 59.

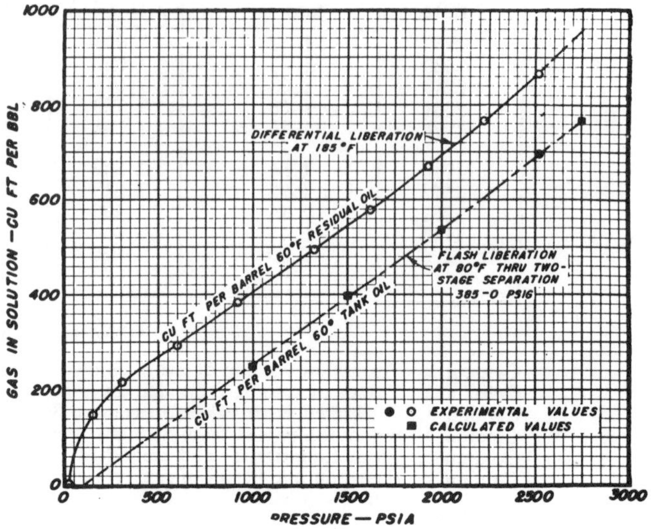

Figure 58—Gas Solubility vs Pressure Obtained by Differential and Flash Processes

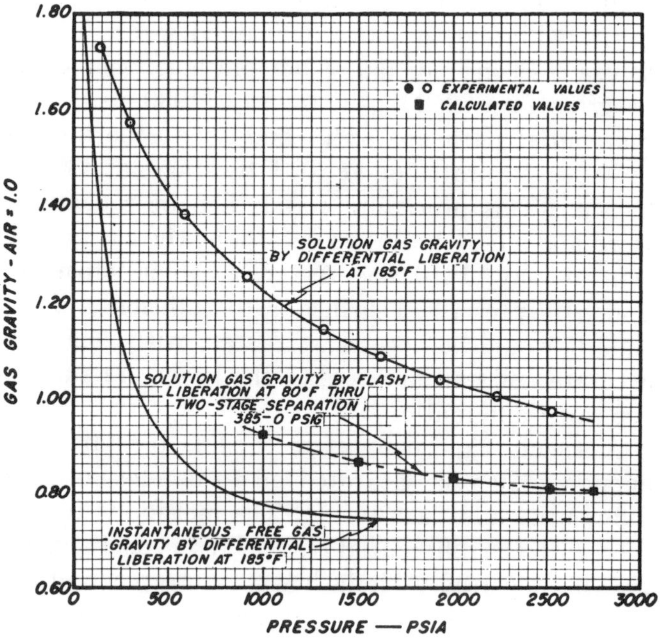

Figure 59—Gravity of Dissolved Gases Obtained by Differential and Flash Processes

The solid black circles in these figures are the experimentally determined values.

The lower curve in Figure 58 represents the surface gas-oil ratio obtained when only saturated reservoir oil is produced from the reservoir. The curve has purposely been extrapolated to zero gas-oil ratio at a pressure near 100 psia. This has been done to illustrate the point that a reservoir oil of low saturation pressure may not produce trap gas or tank vapors when separated under the lower surface temperature.

The gas gravities shown as functions of pressure in Figure 59 illustrate the wide variation that can be obtained when separating gas and liquid by different processes. The lower curve, previously shown in Figure 52, shows the gravity of the gas just coming out of solution in the reservoir oil at 185°F. The upper curve represents the gravity of all the gas still in solution at any pressure when a 185°F differential vaporization process is followed from that pressure to 14.7 psia. The values shown here were obtained by dividing the weight of gas in solution, shown in column 3 of Table 27, by the volume of gas and the constant 0.001222. The intermediate curve represents the total gravity of the solution gas that would be obtained at the surface when only saturated reservoir oil was flashed through the traps. This last value should not be confused with the surface gravity, which might be obtained as a result of simultaneous production of free reservoir gas and saturated reservoir oil.

Pressure-Formation Volume Relations. After the quantity of gas in solution at any particular pressure has been determined, it is a simple matter to calculate the formation volume of the saturated oil as a function of pressure. This calculation is illustrated in Table 33 and makes use of the apparent gas density method outlined in Chapter 4 and a formation volume determined experimentally at some reference pressure.

TABLE 33. METHOD OF CALCULATING FORMATION VOLUME OF SATURATED OIL AS A FUNCTION OF PRESSURE AND GAS SOLUBILITY USING EXPERIMENTAL DATA TO GUIDE CALCULATIONS

1	2	3	4	5	6
		Saturation Pressure—PSIA			
Item	Description	2520	2000	1500	1000
1.	Solubility, cu ft per bbl tank oil	694*	538	395	250
2.	Solution gas gravity, air = 1.0	0.810*	0.830	0.862	0.920
3.	Weight of gas, lb	42.9	34.1	26.0	17.6
4.	Weight tank oil, per bbl, lb	295.0	295.0	295.0	295.0
5.	Total system weight, lb	337.9	329.1	321.0	312.6
6.	Apparent gas density, 14.7 PSIA and 60°F, lb per cu ft	23.9	24.3	24.9	26.0
7.	Pseudo liquid volume of dissolved gas, cu ft	1.795	1.403	1.043	0.676
8.	Pseudo liquid volume of system at 14.7 PSIA and 60°F, cu ft	7.410	7.018	6.658	6.291
9.	Pseudo liquid density of system at 14.7 PSIA and 60°F, lb per cu ft	45.60	46.95	48.20	49.75
10.	Pressure correction to saturation pressure, lb per cu ft	+0.85	+0.65	+0.45	+0.30
11.	Temperature correction from 60°F to 185°F, lb per cu ft	−3.35	−3.25	−3.10	−3.00
12.	Density of system at saturation pressure, lb per cu ft	43.10	44.35	45.55	47.05
13.	Volume of system at saturation pressure, cu ft	7.833	7.425	7.050	6.645
14.	Formation volume of system at saturation pressure, bbl per bbl tank oil	1.395*	1.323	1.255	1.182

*Experimental data

Basis: One barrel tank oil

Notes on Calculation Method

Item 1. From Figure 58
Item 2. From Figure 59
Item 3. Weight $= \frac{\text{cu ft}}{379} \cdot 28.97 \cdot \text{gravity}$
Item 4. From Table 25, tank oil = 36.35° API, or specific gravity = 0.8428 60/60
Item 5. Item 3 plus item 4
Item 6. From Figure 26
Item 7. Item 3 divided by item 6
Item 8. Item 7 plus 5.615
Item 9. Item 5 divided by item 8
Item 10. From Figure 24
Item 11. From Figure 25
Item 12. Item 9 plus item 10 minus item 11
Item 13. Item 5 divided by item 12
Item 14. Item 13 divided by 5.615

Referring to Table 33, items 1, 2, and 14 of column 3 were determined experimentally by flashing a sample of the 2520 psia saturated oil through a particular surface trap arrangement. The derivation of these data is presented in Table 25. The calculation of items 1 through 5 is straightforward, being an expression of the pounds of material produced at the surface per unit barrel of tank oil.

The apparent density that the solution gas exhibits when in solution at the reservoir temperature can be evaluated for the experimental system (column 3) by working backward from the measured formation volume. For example, the reference system consisting of 337.9 lb of material occupies 5.615 · 1.395 = 7.833 cu ft of reservoir space at 2520 psia and 185°F. The density of the system at this pressure and temperature is therefore 43.10 lb per cu ft. Taking into account the compressibility and thermal expansion of the reference system (Figures 24 and 25) results in the evaluation of a pseudo-liquid density of 45.60 lb per cu ft at 14.7 psia and 60°F. This pseudo-liquid density, when divided into the weight of the system, yields a pseudo-liquid volume of $\frac{337.9}{45.60} = 7.410$ cu ft, which is composed of 5.615 cu ft of tank oil and 7.410 − 5.615 = 1.795 cu ft of dissolved gas. The apparent density is obtained from the weight of gas, 42.9 lb, divided by its apparent volume, 1.795 cu ft, or 23.9 lb per cu ft at 14.7 psia and 60°F.

The calculations thus far have been toward evaluating an apparent density for the solution gas obtained from the separation of the reference system by the desired trap operation. After the apparent gas density for the reference system has been calculated, the data can be used to calculate the formation volume of other saturated oils. For example, to obtain the formation volume of the 1500 psia saturated oil, the 23.9 lb per cu ft gas density of the reference system first is plotted on Figure 26, and a curve parallel to the tank oil gravity lines is constructed through the point. The curve so constructed shows that the 0.862 gravity gas obtained from the 1500 psia saturated oil would have an apparent density of 24.9 lb per cu ft.* The volume occupied by the 1500 psia saturated oil at reservoir conditions is calculated by the method described in Chapter 4, which makes use of the weight of solution gas and its apparent density. Example calculations are shown in columns 4, 5, and 6 of Table 33.

* Note that the apparent density values correspond to a tank oil gravity of about 45° API, rather than the 35° API obtained in the laboratory test. Thus, Figure 26 is used only for the relation of apparent density to gas gravity and not for the absolute values of the apparent density.

A comparison is shown in Figure 60 of the relative oil volume obtained by differential liberation at 185°F with the formation volume of the saturated oil as determined from surface trap sets. The solid black circle represents the value of formation volume factor determined experimentally. The solid

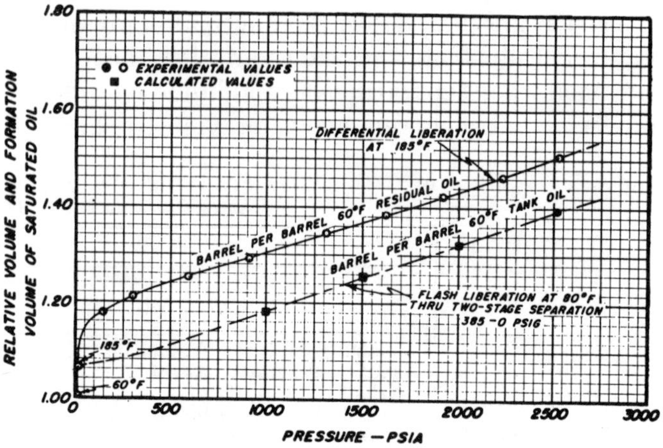

Figure 60—Relative Volumes and Formation Volumes of Saturated Oil vs Pressure

black squares of the lower curve represent the data calculated in Table 33. The left end of the formation volume curve is extrapolated to a value of 1.063 at 14.7 psia and 185°F because of the thermal contraction of 14.7 psia saturated oil in cooling from 185 to 60°F.

Summary of Principal Points

A dissolved gas system has been defined as one that has a critical temperature greater than the temperature at which the system is studied. Because of the particular shape of pressure-temperature phase diagrams of naturally occurring systems, the relation of critical temperature and reservoir temperature is such that only bubble points are encountered at the phase boundary.

The behavior of dissolved gas systems ordinarily is described in terms of the pressure-volume relation of the total system, the pressure-liquid phase volume relation, and the solubility of gas in liquid. Volumetric relation may be expressed either in terms of formation volume or relative volumes. The formation volume defines the volume occupied by the system at elevated pressures and temperatures (formation conditions) per unit volume of 60°F surface tank oil. The relative volume defines the volume of the system at some elevated pressure and temperature per unit volume of saturated oil, or unit volume of 60°F residual oil. Residual oil is the term for the liquid phase that results from a differential vaporization at reservoir temperature. The solubility of gas in oil ordinarily is expressed in units of cu ft of gas, measured at 14.7 psia and 60°F, per barrel of oil. The barrel of oil may be 60°F tank oil, 60°F residual oil, or oil saturated at a specified pressure and temperature.

In most instances, it is of small consequence whether one considers a differential or flash process to prevail in the reservoir. For example, in Figure 50, the volume of liquid phase obtained under flash conditions at 1000 psia and 185°F s shown to be 76.05 cc per 87.92 cc of 2520 psia saturated liquid. In Figure 51, the volume of liquid at the same pressure and temperature, but obtained by a different process, also is 76.05 cc. Thus, in this particular instance, there is no difference in the volumetric behavior of the liquid phase down to 1000 psia, regardless of which process is followed.

Common PVT usage results in some confusion in the meaning of the terms "bubble point pressure" and "saturation pressure." The bubble point has been defined as the state at which an infinitesimal quantity of gas is in equilibrium with a large amount of liquid. As commonly used, the saturation pressure is analogous to bubble point pressure, and its use implies the existence of only a small quantity of gas phase. However, on occasion, the term "saturation pressure" is used to identify the pressure at which vapor and liquid phases are in equilibrium, regardless of the proportion of the phases. The term "bubble point pressure" has a precise meaning and should be used to identify the pressure at which only an infinitesimal gas phase is present.

CHAPTER 8

MATERIAL BALANCES

An important use of PVT data is for evaluating oil and gas reserves. The evaluations can vary in complexity from the simple evaluation of the standard cubic feet of gas per acre-foot of sand to the more complex evaluations that involve the pressure-production performance of reservoirs. It is the purpose of this chapter to discuss the use of PVT data in these more complex types of calculations. It is not proposed to discuss in detail the material balance method, other than to point out a few of the assumptions that are made in deriving material balance equations and to show how these assumptions, in general, will affect the calculated reserves.

The term "material balance," when applied to petroleum reservoir engineering, has come to mean a particular type of calculation. In fact, to many petroleum engineers, it connotes a particular equation proposed by Schilthuis in 1936.[41] As the two words "material" and "balance" suggest, the calculations attempt to evaluate the quantity of "material" present in the reservoir by maintaining a running "volumetric balance" on the material remaining in the reservoir at any particular time during the depletion of the reservoir.

To introduce the more complex reservoir material balance methods, consideration will be given first to a simple idealized system of a liquid and insoluble gas under pressure in a fixed volume container. Following this, a similar system will be discussed in which the gas is soluble in the liquid. Finally, two methods will be outlined for deriving material balance equations applicable to petroleum reservoirs.

Tank Material Balance

The simplest type of material balance equation that can be developed is pictured in terms of a constant-volume tank, initially containing liquid and gas phases. To simplify the problem, it can be assumed that

(1) The gas is insoluble in the liquid at all pressures.

(2) The gas obeys the ideal gas law, $PV = C$ at all pressures.

(3) The tank liquid is incompressible.

(4) The vapor pressure of the liquid is zero.

(5) The tank temperature is equal to the outside atmospheric temperature.

A schematic representation of the tank and the assumed solubility and volumetric relations is shown in Figure 61.

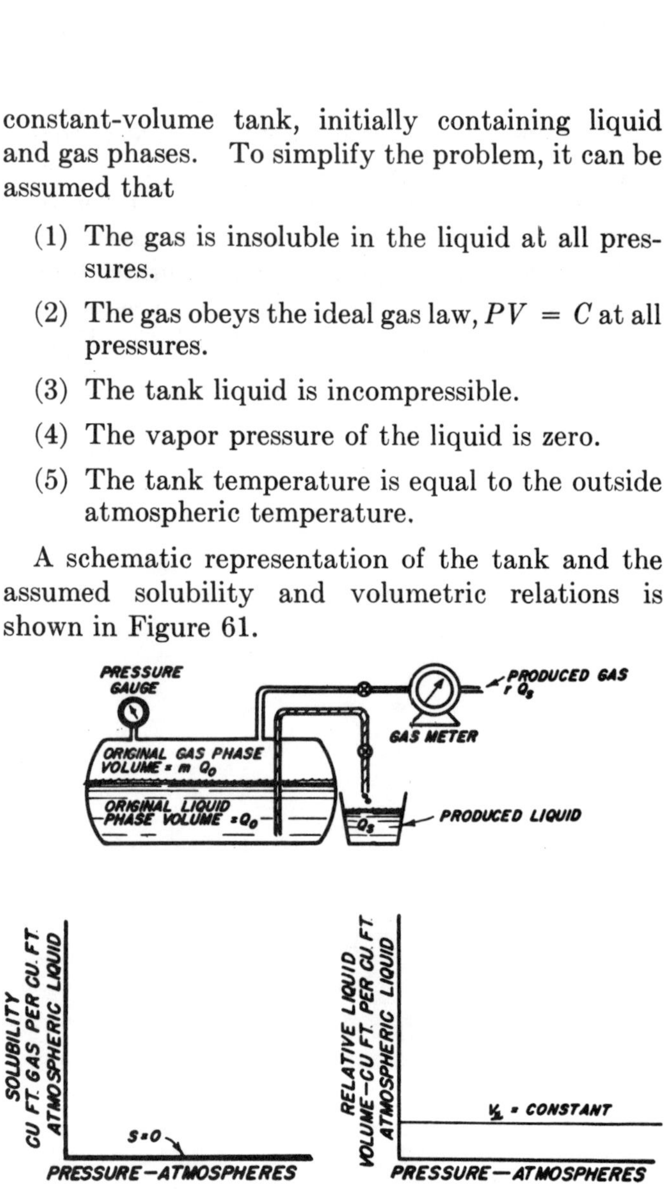

Figure 61—Schematic Representation of Tank and Assumed Solubility and Volumetric Relations Used in Tank Material Balance Derivation

By using the above assumptions, an equation can be developed that relates the pressure in the tank at

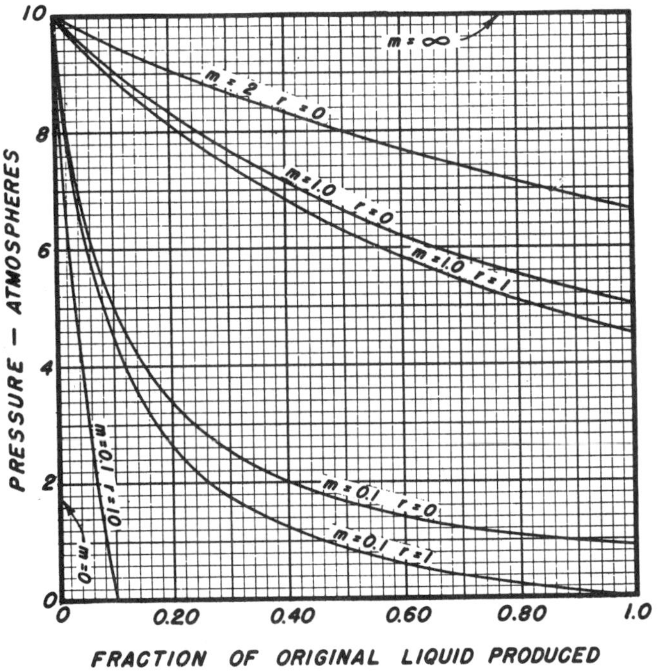

Figure 62—Curves Illustrating the Effect that Size of Gas Cap and Producing Gas-Liquid Ratio Have on the Quantity of Produced Liquid. Calculations Made by Tank Material Balance Method

any time to the quantities of produced gas and liquid.

Let p = tank pressure (atmospheres)
Q_o = volume units of liquid originally in the tank
Q_s = cumulative volume units of liquid produced at time θ
r = cumulative produced gas-liquid ratio measured under atmospheric conditions of pressure and temperature, unit volumes of gas per unit volume of liquid
m = ratio of the original gas volume to the original liquid volume in the tank
Subscript $_o$ indicates original conditions.

Under the original conditions existing in the tank,

Volume of liquid phase $= Q_o$
Volume of gas phase $= mQ_o$

Total tank volume $= Q_o(1 + m)$
Original quantity of gas in the tank $= p_omQ_o$

At subsequent time θ,
Volume occupied by liquid remaining $= (Q_o - Q_s)$
Volume occupied by gas remaining = total tank volume − volume occupied by liquid.
Volume occupied by gas remaining $= Q_o(1 + m) - (Q_o - Q_s)$
Quantity of gas in the tank $= p[Q_o(1 + m) - (Q_o - Q_s)]$

Preparing a material balance on the gaseous material,

Produced gas = original gas − gas remaining
$$rQ_s = p_omQ_o - p[Q_o(1 + m) - (Q_o - Q_s)]$$

which simplifies to

$$Q_o = \frac{Q_s(r + p)}{m(p_o - p)} \tag{50}$$

or the fraction of original liquid that has been produced at time θ is

$$\frac{Q_s}{Q_o} = \frac{m(p_o - p)}{r + p} \tag{51}$$

Equation (50) is a basic type of material balance equation that evaluates the original quantity of liquid in the tank as a function of two pressures, the cumulative produced gas-liquid ratio, and the ratio of gas and liquid volumes originally present in the container. Two unknowns, Q_o and m, and only a single equation require that a second equation be developed to obtain specific values of Q_o.

A plot of Equation (51) is shown in Figure 62. The fraction of the liquid recovered at any pressure is shown for several values of m and r.

Pop Bottle Material Balance

A slightly more complicated material balance equation results from introducing solubility of gas in the liquid, and the consequent expansion of the liquid phase. The following has been termed the "pop bottle" material balance equation, as the producing process is visualized best as occurring in the familiar charged soda water bottle. A schematic representation of the bottle and of the assumed solubility and volumetric conditions is

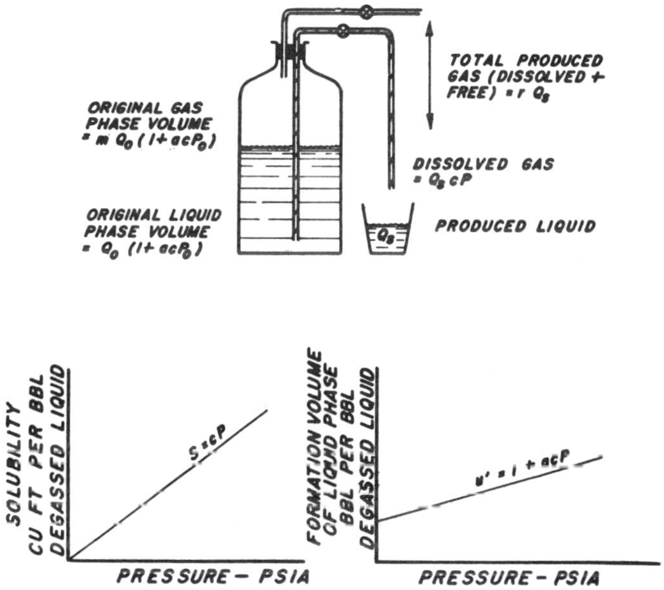

Figure 63—Schematic Representation of Pop Bottle and Assumed Solubility and Volumetric Relations Used in Pop Bottle Material Balance Derivation

shown in Figure 63. The assumptions used in the derivation are as follows:

(1) Solubility of the gas in the liquid is proportional to the absolute pressure (Henry's Law).

(2) Free gas in the bottle obeys the ideal gas law, $PV = C$.

(3) The liquid phase is incompressible.

(4) Vapor pressure of the liquid is zero.

(5) The bottle temperature and the atmospheric temperature are the same (assumed 60°F in example).

(6) Solution of each increment of gas causes an equal volumetric increase of the liquid phase.

The linear solubility-pressure relation can be expressed in petroleum engineering units as follows:

$$S = cP$$

where S = solubility, cu ft of gas per bbl of "degassed" liquid
P = absolute pressure (psia)
c = solubility constant, cu ft per bbl per psi

The pressure-volume relationship of the liquid phase can be written

$$u' = 1 + acP$$

where u' = volume of liquid phase containing dissolved gas, per unit volume of "degassed" liquid
a = expansion coefficient of the liquid phase caused by the solution of one cubic foot of gas in a unit barrel of "degassed" liquid

Letting Q_o be the number of barrels of "degassed" liquid originally in the bottle,

Volume (bbl) of liquid phase $= Q_o(1 + acP_o)$
Volume (bbl) of gas phase $= mQ_o(1 + acP_o)$

Total original volume (bbl) $= Q_o(1 + acP_o)(1 + m)$
The standard cubic feet of free gas originally present

$$= \frac{5.615}{14.7} P_o m Q_o(1 + acP_o)$$

The cubic feet of dissolved gas originally present $= Q_o c P_o$
Total cubic feet of gas originally in the system

$$= \frac{5.615}{14.7} P_o m Q_o(1 + acP_o) + Q_o c P_o$$

At any subsequent time, θ, the volume of the phases can be expressed as follows:

Volume (bbl) occupied by liquid phase $= (Q_o - Q_s)(1 + acP)$
Volume (bbl) occupied by free gas = total volume − volume occupied by liquid phase
Volume (bbl) occupied by free gas

$$= Q_o(1 + acP_o)(1 + m) - (Q_o - Q_s)(1 + acP)$$

Standard cu ft of free gas

$$= \frac{5.615}{14.7} P[Q_o(1 + acP_o)(1 + m) - (Q_o - Q_s)(1 + acP)]$$

Cu ft of dissolved gas $= (Q_o - Q_s)cP$

As the cumulative produced gas is the difference between the original gas present and the gas remaining at time θ,

$$rQ_s = \frac{5.615}{14.7} P_o m Q_o(1 + acP_o) + Q_o c P_o - \frac{5.615}{14.7} P[Q_o(1 + acP_o)(1 + m) - (Q_o - Q_s)(1 + acP)] - (Q_o - Q_s)cP$$

which simplifies to

$$Q_o = \frac{Q_s[r + (0.382 - c)P + 0.382acP^2]}{0.382m[(1 + acP_o)(P_o - P)] + cP_o - c(1 + 0.382aP_o)P + 0.382acP^2} \quad (52)$$

or, in terms of the fraction of original liquid that has been produced, $\frac{Q_s}{Q_o}$

$$\frac{Q_s}{Q_o} = \frac{0.382m[(1 + acP_o)(P_o - P)] + cP_o - c(1 + 0.382aP_o)P + 0.382acP^2}{r + (0.382 - c)P + 0.382acP^2} \quad (53)$$

As would be expected, Equation (53) is similar in form to Equation (51). Equation (53) is quadratic in P, but is linear with respect to $\frac{Q_s}{Q_o}$. It is solved easily for specific values of m, a, c, and r, as illustrated in Table 34 and Figure 64.

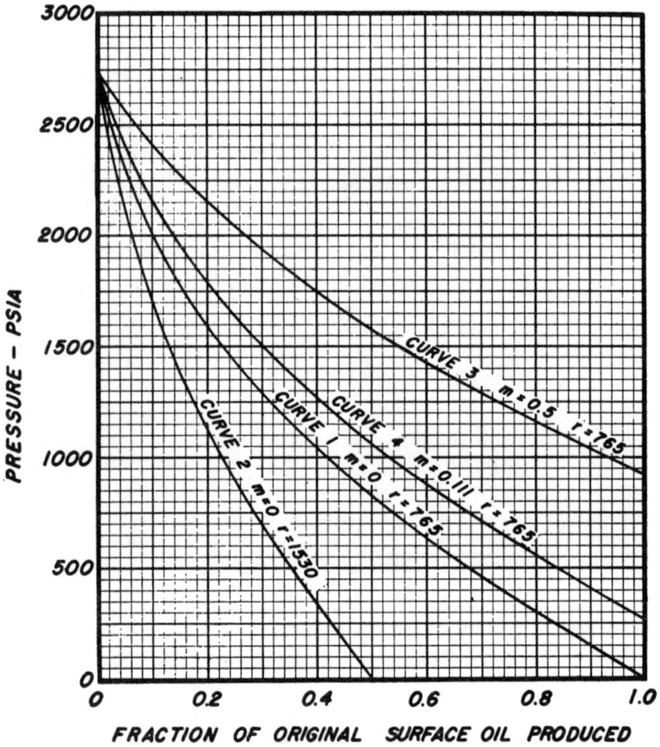

Figure 64—Curves Illustrating the Effect that Size of Gas Cap and Producing Gas-Oil Ratio Have on the Quantity of Produced Oil. Calculations Made by Pop Bottle Material Balance Method

TABLE 34. EXAMPLE APPLICATION OF POP BOTTLE MATERIAL BALANCE METHOD TO RESERVOIR BEHAVIOR

Problem: To calculate reservoir pressure as a function of the quantity of produced surface oil, making use of the following assumed fluid properties:

Original pressure	2750 psia
Original gas solubility	765 cu ft per bbl tank oil
Formation volume of original saturated liquid phase	1.420 bbl per bbl tank oil

Solution:
From Equation (53)

$$\frac{Q_s}{Q_o} = \frac{0.382m[(1 + acP_o)(P_o - P)] + cP_o - c(1 + 0.382aP_o) + 0.382acP^2}{r + (0.382 - c)P + 0.382acP^2}$$

$$c = \frac{765}{2750} = 0.278 \qquad cP_o = 765$$

$$a = \frac{0.420}{765} = 5.49(10^{-4}) \qquad aP_o = 5.49(10^{-4})2750 = 1.51$$

$$acP_o = 0.420$$

Case 1. No free gas at start. Producing ratio equal to original solution ratio.

$$m = 0 \qquad r = 765$$

$$\frac{Q_s}{Q_o} = \frac{765 - 0.438P + 0.582(10^{-4})P^2}{765 + 0.104P + 0.582(10^{-4})P^2}$$ Curve 1, Figure 64 is a plot of this equation.

Case 2. No free gas at start. Producing ratio equal to twice original solution ratio.

$$m = 0 \qquad r = 1530$$

$$\frac{Q_s}{Q_o} = \frac{765 - 0.438P + 0.582(10^{-4})P^2}{1530 + 0.104P + 0.582(10^{-4})P^2}$$ Curve 2, Figure 64 is a plot of this equation.

Case 3. One-third free gas at start. Producing ratio equal to original solution ratio.

$$m = 0.5 \qquad r = 765$$

$$\frac{Q_s}{Q_o} = \frac{1510 - 0.709P + 0.582(10^{-4})P^2}{765 + 0.104P + 0.582(10^{-4})P^2}$$ Curve 3, Figure 64 is a plot of this equation.

Case 4. 10% free gas at start. Producing ratio equal to original ratio.

$$m = 0.111 \qquad r = 765$$

$$\frac{Q_s}{Q_o} = \frac{931 - 0.500P + 0.582(10^{-4})P^2}{765 + 0.104P + 0.582(10^{-4})P^2}$$ Curve 4, Figure 64 is a plot of this equation.

Equation (53) is equivalent to the material balance equation published by Coleman, Wilde, and Moore[43] in 1930. The assumptions of ideal gas behavior, linear gas solubility and liquid expansion with pressure, and constant reservoir volume limit its usefulness in petroleum reservoir problems. The equation is of interest, however, because the authors mentioned above first used it to illustrate the possibility of obtaining greater oil recovery from a reservoir by controlling the produced gas-oil ratio. This can be seen by comparing the values at equal pressures shown by curves 1 and 2, Figure 64.

Reservoir Material Balance, I

The third and fourth methods of deriving a material balance equation are termed "reservoir" material balance methods because they take into account the physical properties of the fluids as they exist in oil reservoirs. The methods make use of relations measured in the laboratory, rather than the straight-line relations of gas solubility and liquid phase volume to pressure illustrated in Figure 63 and the assumption that reservoir gases follow ideal gas behavior. In addition, a factor is introduced into the equation to allow for the change in reservoir hydrocarbon volume caused by encroaching waters. The first of the reservoir material balance methods is similar to the pop bottle method, outlined previously. The second is a much simpler derivation than the first and is based on a combination of the methods proposed by Schilthuis[41] and Cook.[44]

A schematic representation of an idealized reservoir is shown in Figure 65. Pressure-solubility relations of gas and formation volumes of the reservoir liquid and gas phases also are shown. The formation volume and solubility are expressed in terms of a unit barrel of surface tank oil and are, therefore, functions of the method by which the gas and oil are separated at the surface. The following two assumptions enter into the derivation.

(1) The reservoir unit is sufficiently small to permit the same pressure to exist throughout the unit at any time.

(2) The degree of equilibrium achieved in PVT tests is the same as in the reservoir.

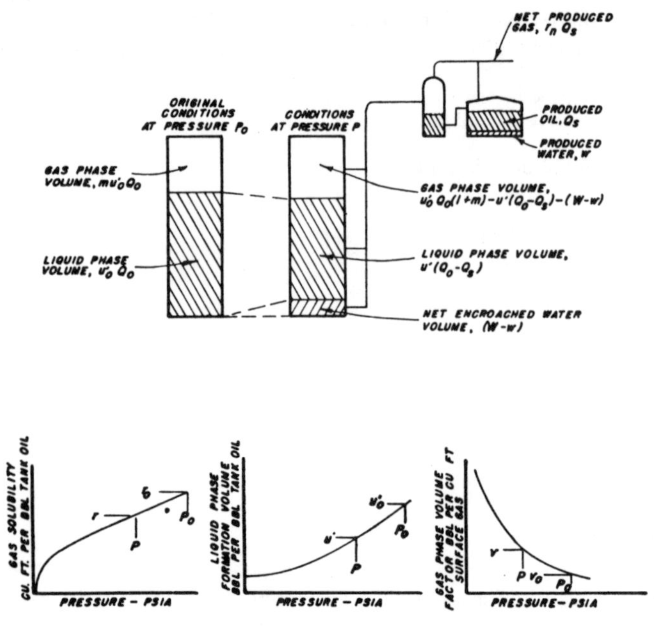

Figure 65—Schematic Representation of Idealized Reservoir and Assumed Solubility and Volumetric Relations Used in Reservoir Material Balance I Derivation

When the same approach is used as in the pop bottle derivation, the original phase volumes and gas quantities at pressure P_o are as follows:

Liquid phase, volume (bbl) $= u'_o Q_o$
Gas phase volume (bbl) $= mu'_o Q_o$

Total original volume (bbl) $= u'_o Q_o (1 + m)$

where u'_o = formation volume of original liquid phase, bbl per bbl of tank oil
Q_o = bbl of tank oil originally present

Standard cu ft of free gas originally present $= mu'_o Q_o \frac{1}{v_o}$

Standard cu ft of dissolved gas originally present $= r_o Q_o$

Total gas (standard cu ft) $= mu'_o Q_o \frac{1}{v_o} + r_o Q_o$

where v_o = volume, in bbl, of one cubic foot of free gas when under reservoir pressure and temperature conditions

$$v_o = \frac{1}{5.615} \cdot \frac{14.7}{P_o} \cdot \frac{T}{520} \cdot Z_o$$

$$= 5.03(10^{-3}) \frac{TZ_o}{P_o}$$

r_o = original dissolved gas-oil ratio, cu ft of surface gas per bbl of tank oil

At any subsequent time, θ, and at reservoir pressure, P, the volume of the phases can be expressed as follows:

Liquid phase volume (bbl) $= u'(Q_o - Q_s)$
Free gas phase volume = total volume − liquid phase volume − net water encroachment

Free gas phase volume (bbl)
$$= u'_o Q_o (1 + m) - u'(Q_o - Q_s) - (W - w)$$

where u' is the formation volume of the liquid phase
W = cumulative bbl of water encroached into reservoir at time θ
w = cumulative bbl of produced water at time θ

The cu ft of free gas

$$= \frac{1}{v} [u'_o Q_o (1 + m) - u'(Q_o - Q_s) - (W - w)]$$

The cu ft of dissolved gas $= (Q_o - Q_s) r$

Striking a material balance of the "gas component," we have:

Net produced quantity of gas = original gas − quantity remaining at time θ.

$$r_n Q_s = mu'_o Q_o \frac{1}{v_o} + r_o Q_o - \frac{1}{v} \Big[u'_o Q_o (1 + m) - u'(Q_o - Q_s) - (W - w) \Big] - (Q_o - Q_s) r$$

which simplifies to

$$Q_o = \frac{Q_s[u' + (r_n - r)v] - (W - w)}{u' - u'_o + mu'_o \left(\frac{v}{v_o} - 1\right) + (r_o - r)v} \qquad (54)$$

where r_n = cumulative net produced gas-oil ratio, cu ft per bbl of tank oil
r = dissolved gas-oil ratio, cu ft per bbl of tank oil

It is important to note that the terms Q_o, Q_s, r, and u' are results of the type of flash separation conditions employed at the surface to separate gas and oil. The last two factors would be obtained from curves similar to the lower curves in Figures 58 and 60. Further, it should be realized that to say a reservoir originally contains Q_o barrels of tank oil implies that all the reservoir fluids will be processed through a particular surface trapping arrangement. If, at some time during the depletion of a reservoir, the surface trapping conditions are changed, it then becomes necessary to reevaluate Q_o, u', and r in terms of the new separation process.

Equation (54) considers the saturated liquid and free gas as two separate hydrocarbon quantities in the reservoir. It does not differentiate completely between produced gas originating from the gas cap and produced gas that was originally dissolved in the liquid phase. The term $(r_o - r)$ in the denominator of Equation (54) represents the volume of free gas at pressure P which was originally dissolved in oil at pressure P_o. The v in this term should be obtained by using a compressibility factor, Z, that is representative of gas coming out of solution. The middle term in the denominator, $mu'_o \left(\frac{v}{v_o} - 1\right)$, clearly represents the expansion of gas cap material. If the gas in the gas cap is of a composition different from gas originating in the oil band, or if the pressure in the gas cap is materially different from that in the oil band, the v of the middle term will not be the same as the v in the third term. However, as the middle term v's appear as a ratio, the effect of differences in values caused by composition is minimized. The v appearing in the numerator cannot be defined rigorously because gas in the net produced gas-oil ratio, r_n, is not specified as to origin.

Reservoir Material Balance, II

A much simpler derivation of a material balance equation can be made by evaluating the expansion of the original liquid phase and the free gas phase as reservoir pressure is changed. Referring to Figure 66, let it be assumed that we are observing the

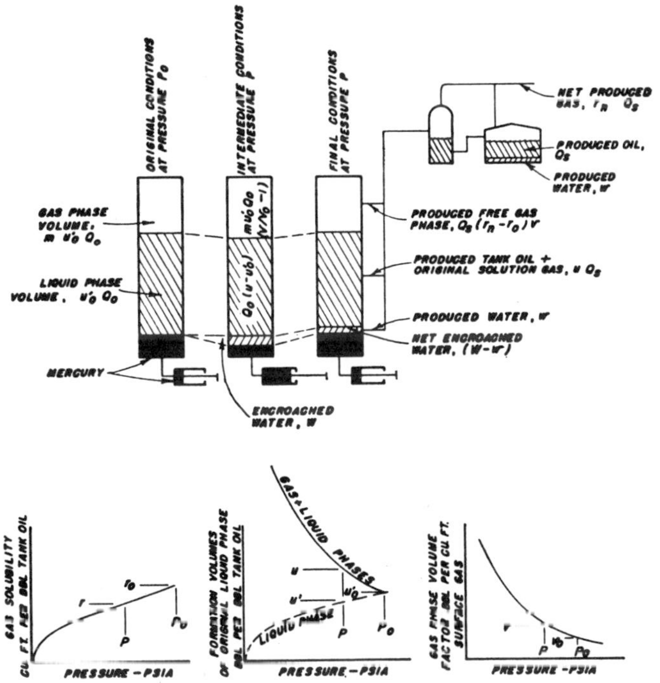

Figure 66—Schematic Representation of Idealize Reservoir and Assumed Solubility and Volumetric Relations Used in Reservoir Material Balance II Derivation

behavior of an idealized segment of a reservoir that contains a saturated oil and a gas cap. For the purpose of describing the process, let us assume also that an inert liquid, such as mercury, can be injected freely into or withdrawn from the bottom of the reservoir.

At the original pressure P_o, the volumes of the two hydrocarbon phases are expressed as

$$\text{Volume of liquid phase (bbl)} = u'_oQ_o$$
$$\text{Volume of free gas phase (bbl)} = mu'_oQ_o$$

If a quantity of mercury is withdrawn from the segment, causing the pressure to drop from P_o to P, the volumes occupied by the original phases will be

$$\text{Volume of original liquid phase (bbl)} = uQ_o$$

$$\text{Volume of free gas phase (bbl)} = mu'_oQ_o\frac{v}{v_o}$$

where u is the **two-phase** formation volume of what originally was saturated liquid. (Note the difference here in the use of u and u' of the previous method.)

The sum of the expansions of the original free gas phase and the original liquid phase plus the volume of any encroached water, W, caused to flow into the segment by the reduction in pressure is, of course, equal to the volume of mercury withdrawn. Mathematically, this can be represented as

$$\text{Total expansion} = mu'_oQ_o\left(\frac{v}{v_o} - 1\right) + Q_o(u - u'_o) + W$$

Let us now assume that gas, oil, and water are produced from the segment while the pressure remains constant at P. This can be done by reinjecting the mercury removed previously. Obviously, to keep the pressure constant, the volume of reinjected mercury will be equal to the sum of the produced gas, oil, and water volumes, all measured at pressure P. These quantities are expressed as follows:

Volume of produced water (bbl) $= w$
Volume of produced tank oil and gas originally in solution at P_o in that tank oil (bbl) $= uQ_s$
Volume of free gas produced is equal to the difference between the net produced gas and the gas originally dissolved in the produced tank oil. At pressure P, volume of produced free gas $= Q_s(r_n - r_o)v$
Thus, the volumetric contraction of the system $= Q_s(r_n - r_o)v + uQ_s + w$

As the expansion of the idealized reservoir in the first stage of the process must be equal to the contraction in volume in the second,

$$mu'_oQ_o\left(\frac{v}{v_o} - 1\right) + Q_o(u - u'_o) + W = Q_s(r_n - r_o)v + uQ_s + w$$

or, solving for Q_o

$$Q_o = \frac{Q_s[u + (r_n - r_o)v] - (W - w)}{u - u'_o + mu'_o\left(\frac{v}{v_o} - 1\right)} \tag{55}$$

Equation (55) is the familiar Schilthuis material balance equation. This equation also can be obtained from Equation (54) by substituting $u - (r_o - r)v$ for the liquid phase formation volume, u', of Equation (54). The difficulty of rigorously defining the v terms is the same as that expressed in the previous section.

For many purposes, it is advantageous to separate the net produced gas-oil, r_n, into its two components: the produced gas-oil ratio, r_p, and the injected gas-oil ratio, r_i, giving

$$Q_o = \frac{Q_s[u + (r_p - r_o)v - r_i v_i] - (W - w)}{u - u'_o + mu'_o\left(\frac{v}{v_o} - 1\right)} \quad (56)$$

If G is the cubic feet of surface gas injected into the reservoir during a gas injection program, G can be expressed in terms of an injected gas-oil ratio. Thus,

$$G = Q_s r_i$$

then

$$Q_o = \frac{Q_s[u + (r_p - r_o)v] - Gv_i - (W - w)}{u - u'_o + mu'_o\left(\frac{v}{v_o} - 1\right)} \quad (57)$$

where v_i = volume factor of the injected gas, bbl at reservoir conditions per cu ft of surface gas

Where applicable, Equation (57) has a small advantage over Equation (55) because of its better representation of the compressibilities of the separate gases.

Application of Material Balance Equations, Douglas Field

To this point, we have considered four methods of developing material balance equations. The tank material balance equation has no application to reservoir problems, nor does the pop bottle equation. The method outlined as Reservoir Method I has applicability to reservoir problems; in fact, it has been used by Tarner[42] in presenting a method of calculating reservoir pressure-production behavior. By far the most used material balance equation is that developed by Schilthuis (Equation 55), and it is on the basis of this equation that the application of material balance equations to reservoir engineering problems will be discussed.

It should be clear by now that the material balance equation is a mathematical expression of gas, oil, and water volumes originally present in a petroleum reservoir coupled with the movements of these or similar phases into or out of the reservoir. The equations say nothing regarding movement of fluid volumes within the limits of the reservoir. To put it another way, in developing Equation (55), an imaginary boundary was placed around a volume of earth and an attempt made to estimate the quantity and type of fluid contained within that volume by observing the manner in which pressure changes within the confines of the boundary as fluids are moved across it. Unfortunately, it is possible to observe and measure only the fluids flowing across the boundary face at the earth's surface; fluids flowing across it at other locations can be evaluated only by indirect calculations.

Equation (55) contains three unknown fluid quantities: the original tank oil in place, Q_o; the volume of water entering the reservoir, W; and the volume of the original gas cap in the reservoir mu'_oQ_o. Obviously, it is impossible to arrive at a unique solution to the material balance equation without two additional independent equations. The need for additional independent equations has been well illustrated by Woods and Muskat.[45] They show that the pressure-production history of the Jones Sand Reservoir, Schuler Pool, can be duplicated with equal precision by calculations that assume any of the following conditions:

(a) Original tank oil in place, Q_o = 116(10^6) bbl
No original gas cap
No water encroachment

(b) Original tank oil in place, Q_o, = 93.9(10^6) bbl
No original gas cap
Water encroachment, W, = 15.7(10^6) bbl

(c) Original tank oil in place, Q_o, = 75.3(10^6) bbl
Original free gas = 10.7(10^9) cu ft
Water encroachment, W, = 14.0 (10^6) bbl

Thus, if one were forced to evaluate the original reserves of the Schuler Pool solely by use of a material balance equation and the pressure-production history of the pool, the answer might vary by 40 million bbl of oil and by 10.7 billion cu ft of gas.

The remainder of this chapter will be used to give an example material balance for the Douglas Field and to evaluate water influx into the reservoir.

The example also will serve as a basis for a general discussion of several factors that enter into material balance calculations. The properties of the Douglas reservoir liquid have been presented previously in Chapter 7; the field production and results of periodic pressure surveys are presented in Table 35 and shown graphically in Figure 67. These data show that from April 1, 1943, to January 1, 1950, the field produced 19.5 million bbl of tank oil, 27.0 billion cu ft of gas, and 4.5 million bbl of water. In so doing, the average reservoir pressure dropped from 2750 to 1720 psia, although the rate of pressure decline was lowered materially after April, 1948, by the injection of 1250 MCF of gas a day.

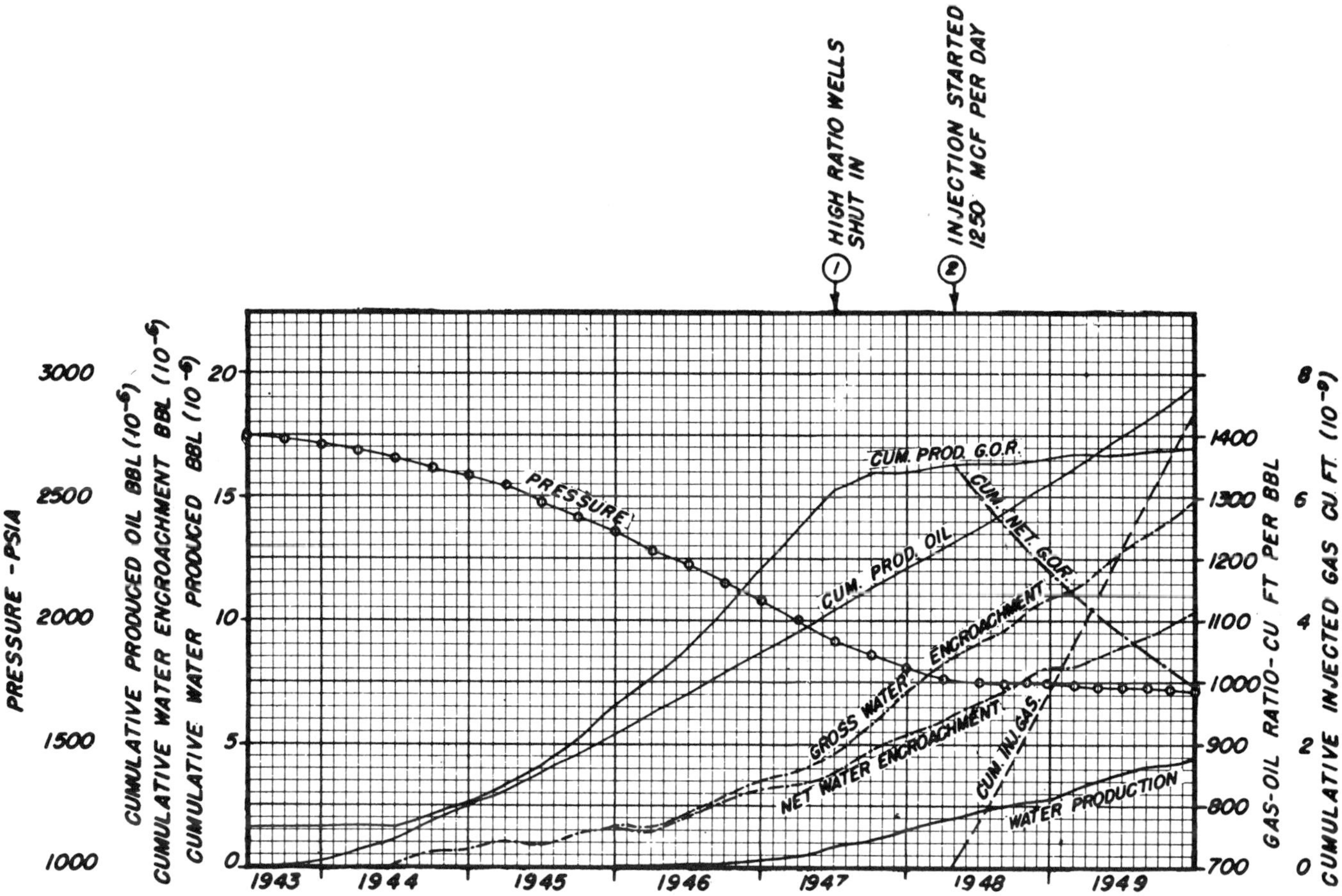

Figure 67—Pressure-Production History of the Douglas Field

Estimation of Original Hydrocarbons in Place. To evaluate the water influx term, W, in Equation (57), it is necessary to evaluate independently the original tank oil in place, Q_o, and the ratio of free gas phase volume to liquid phase volume, m. These values ordinarily are obtained from a study of electric logs of drilled wells, the results of core analysis, and the formation volume of the original reservoir liquid.

In the Douglas Field, the reservoir sand volume has been estimated to be 96,000 acre-feet. This value was obtained from electric logs by calculating the net volume of sand above the water table that was available to hydrocarbons. Core analysis results were also of assistance in evaluating the net sand volume.

The results of porosity, fluid saturations, and interstitial water tests made on cores from selected wells were used with the above sand volume to determine the volumes of hydrocarbons initially in the field. From an average porosity of the sand, 21.4 per cent, and an average interstitial water of 32 per cent, the total hydrocarbon volume was calculated to be

$$\frac{96{,}000 \cdot 43{,}560 \cdot 0.214(1 - 0.32)}{5.615} = 108.5(10^6) \text{ bbl.}$$

Core analysis results indicated that 7.0 per cent of the hydrocarbon volume was occupied by a free gas phase; therefore, the volume of saturated oil amounted to $0.93 \cdot 108.5(10^6) = 100.9(10^6)$ bbl. The volume occupied by free gas amounted to $0.07 \cdot 108.5(10^6) = 7.6(10^6)$ bbl.

From the data presented in Figure 60, the formation volume of the original saturated liquid is 1.422 bbl per bbl of tank oil. Therefore, the original tank oil in place amounted to $\frac{100.9(10^6)}{1.422} = 71.0(10^6)$ bbl. If it is assumed that the compressibility of the original free gas is equivalent to that of gas coming out of solution at 2750 psia, the gas volume factor, v_o, relating the barrels of reservoir space occupied by one cubic foot of surface is

$$\frac{1}{5.615} \cdot \frac{14.7}{2750} \cdot \frac{645}{520} \cdot 0.797 = 0.00094 \text{ bbl per cu ft}$$

(compressibility factor from Figure 52). The quantity of free gas in the reservoir, measured under surface conditions, therefore amounted to

$$\frac{7.6(10^6)}{0.00094} = 8.09(10^9) \text{ standard cu ft.}$$

Properties of the Original Hydrocarbons in Place. Results of PVT tests on a bottom-hole liquid sample that had a bubble point pressure of 2520 psia at 185°F are given in Chapter 7. Methods were outlined by which the data could be extended to give the properties of a liquid having a saturation pressure of 2750 psia at 185°F, the original reservoir liquid in the present example.

TABLE 35. SAMPLE MATERIAL BALANCE CALCULATION, DOUGLAS FIELD

$Q_o = 71(10^6)$ Bbl $\quad u'_o = 1.422$ Bbl per Bbl Tank Oil

$v_o = 0.000940$ Bbl per Cu Ft

1	2	3	4	5	6	7	8	9
End of Prior Date	Average Pressure PSIA	Cumulative Produced Oil, Q, bbl. (10^{-6})	Cumulative Produced Gas cu ft (10^{-9})	Cumulative Produced Water, w bbl. (10^{-6})	Cumulative Injected Gas, G cu ft (10^{-6})	Cumulative Produced Gas-Oil Ratio, r_p cu ft per bbl	Formation Volume Factor, u bbl per bbl tank oil	Produced Free Gas Volume Factor, v bbl per cu ft (10^3)
1943								
June	2750	0	0			765	1.4220	0.940
Sept.	2735	0.220	0.168			765	1.4247	0.947
Dec.	2720	0.310	0.235			765	1.4275	0.950
1944								
March	2690	0.652	0.499			765	1.4333	0.966
June	2655	1.202	0.926			770	1.4402	0.980
Sept.	2620	1.952	1.534			785	1.4474	0.990
Dec.	2585	2.500	2.012			805	1.4562	1.010
1945								
March	2550	3.152	2.633			835	1.4626	1.025
June	2475	3.778	3.266			865	1.4803	1.055
Sept.	2420	4.652	4.245	0		910	1.4935	1.082
Dec.	2360	5.357	5.157	0.051		962	1.510	1.110
1946								
March	2275	6.205	6.257	0.097		1007	1.536	1.155
June	2225	7.030	7.415	0.152		1055	1.553	1.180
Sept.	2150	7.950	8.891	0.227		1117	1.577	1.225
Dec.	2085	8.750	10.375	0.302		1185	1.602	1.270
1947								
March	2000	9.502	11.873	0.452		1248	1.635	1.325
June	1920	10.350	13.570	0.827		1310	1.670	1.390
Sept.	1860	11.278	15.235	1.072		1338	1.699	1.440
Dec.	1810	12.158	16.322	1.563		1343	1.726	1.485
1948								
March	1770	12.852	17.345	1.802		1350	1.750	1.520
April	....	13.205	17.870	.....	0	1352	.....	.....
June	1750	13.780	18.615	2.354	0.71	1353	1.763	1.540
Aug.	1745	14.329	10.356	2.502	1.35	1353	1.765	1.547
Oct.	1745	14.995	20.371	2.675	2.10	1358	1.765	1.547
Dec.	1740	15.652	21.310	2.773	2.82	1362	1.769	1.550
1949								
Feb.	1740	16.150	22.075	3.182	3.60	1366	1.769	1.550
April	1735	16.850	23.078	3.557	4.38	1370	1.772	1.553
June	1735	17.495	23.980	3.852	5.15	1371	1.772	1.553
Aug.	1730	18.150	24.975	4.073	5.95	1375	1.775	1.560
Oct.	1725	18.853	26.003	4.253	6.72	1378	1.778	1.565
Dec.	1720	19.554	27.000	4.473	7.48	1381	1.781	1.570

Two-phase formation volume factors of the original reservoir liquid are listed in column 8, Table 35, and are shown as a function of pressure in Figure 68. These values were obtained by calculation from the smoothed "Y" curve function presented in column 7, Table 30, and shown in Figure 68. Use of the "Y" function greatly simplifies the work of tabulating values of u at pressures less than 2750 psia. For example, at 2550 psia,

$$Y = 2.747; \qquad u = 1.422 + \frac{1.422}{2.747}\left(\frac{2750 - 2550}{2550}\right) = 1.4626.$$

The gas volume factors, v, shown in Figure 68 were calculated by using the compressibility factors shown in Figure 52. It was assumed that the gas cap material was of comparable composition to the gas coming from solution in the reservoir oil when the compressibility factors were used to calculate the values of v shown in column 10, Table 35. The gas volume factors in column 12 have been calculated on the basis of 0.60 gravity injection gas. In this instance the compressibility factors were obtained through use of pseudo-critical constants for a 0.60 gravity gas and the compressibility factor chart shown in Figure 15.

Relative Phase Volumes. The ratio of the reservoir volume originally occupied by free gas phase to the volume occupied by liquid can be evaluated by several methods. Core analysis results showed that 7.0 per cent of the reservoir sand volume contained free gas. Consequently, the value of m is $\dfrac{0.07}{1 - 0.07} = 0.075$.

When a free gas phase does not exist in the reservoir, the value of m in Equation (57) is zero. However, the fact that undersaturated fluid is produced

TABLE 35—(*Continued*)

10	11	12	13	14	15	16	17
$u+(r_p-r_o)v$ bbl per bbl tank oil	$Q_s[u+(r_p-r_o)v]$ bbl (10^{-6})	Injected Gas Volume Factor, v_1 bbl per cu ft (10^3)	Gv, bbl (10^{-6})	$(u-u'_o)+mu'_o\left(\frac{v}{v_o}-1\right)$ bbl per bbl tank oil	$Q_o\left[u-u'_o+mu'_o\left(\frac{v}{v_o}-1\right)\right]$ bbl (10^{-6})	Gross Water Encroachment W bbl (10^{-6})	Net Water Encroachment (W−w) bbl (10^{-6})
1.4220	0			0	0		
1.4247	0.313			0.0035	0.248	0.065	0.065
1.4275	0.443			0.0066	0.469	−0.026	0.026
1.4333	0.935			0.0142	1.008	−0.073	−0.073
1.4451	1.737			0.0227	1.612	0.125	0.125
1.4672	2.864			0.0311	2.208	0.656	0.656
1.4966	3.742			0.0421	2.989	0.753	0.753
1.5344	4.836			0.0502	3.564	1.272	1.272
1.5858	5.991			0.0713	5.062	0.929	0.929
1.6504	7.678			0.0876	6.220	1.458	1.458
1.729	9.262			0.1073	7.618	1.695	1.644
1.816	11.268			0.1384	9.826	1.539	1.442
1.895	13.322			0.1582	11.232	2.242	2.090
2.008	15.964			0.1873	13.298	2.893	2.666
2.135	18.681			0.2174	15.435	3.548	3.246
2.275	21.617			0.2567	18.226	3.843	3.391
2.414	24.985			0.2991	21.236	4.576	3.749
2.524	28.466			0.3337	23.693	5.825	4.753
2.584	31.420			0.3658	25.972	7.011	5.448
2.639	33.916			0.3939	27.960	7.758	5.956
.....	.....		0	......	......	.....	.....
2.669	36.779	1.665	1.182	0.4091	29.046	8.903	6.549
2.675	38.330	1.670	2.254	0.4119	29.245	9.333	6.831
2.683	40.231	1.670	3.507	0.4119	29.245	10.154	7.479
2.694	42.166	1.675	4.723	0.4162	29.550	10.666	7.893
2.701	43.621	1.675	6.030	0.4162	29.550	11.223	8.041
2.712	45.697	1.680	7.358	0.4195	29.785	12.111	8.554
2.713	47.464	1.680	8.652	0.4195	29.785	12.879	9.027
2.727	49.495	1.685	10.026	0.4233	30.054	13.488	9.415
2.737	51.601	1.690	11.357	0.4269	30.310	14.187	9.934
2.748	53.734	1.695	12.679	0.4305	30.565	14.963	10.490

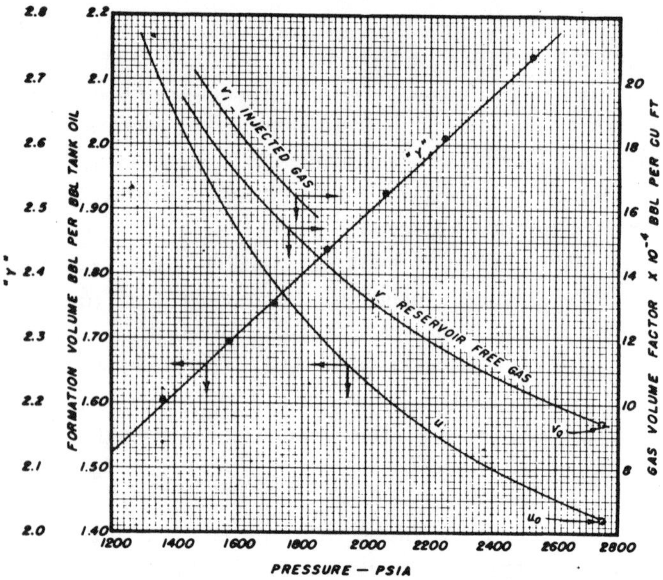

Figure 68—Formation Volume and Gas Volume Factors Used in Illustrative Material Balance Problem

from several wells in a reservoir is not always proof that a gas cap does not exist at some other location in the reservoir. Reservoirs having large vertical sections can have gas caps in contact with an oil band which progressively becomes more undersaturated at points farther removed from the gas cap. The importance of obtaining the best possible value of m is evident from the fact that the compressibility of free gas at, say, 3000 psia and 200°F is of the order of $330(10^{-6})$ vol per vol per psi, while the compressibility of undersaturated oils is of the order of $15(10^{-6})$ vol per vol per psi. Thus, with a 20-fold ratio in compressibilities, a small gas cap can play an important part in the pressure behavior of the reservoir.

The Water Encroachment. In the example calculation shown in Table 35, the gross water encroachment, W, was the unknown factor to be evaluated by use of Equation (57). The other two variables in Equation 57, Q_o and m, were evaluated from geologic and core analysis data.

During the early portion of a field's history it is difficult, if not impossible, to obtain from geologic and core data a good estimate of the original oil in place and the possible existence of a free gas phase. As the number of wells entering the reservoir increases, the estimation of original reserves becomes more exact and, consequently, the water encroachment term can be evaluated with greater precision.

Several individuals have proposed methods independent of material balance calculations to evaluate the water encroachment into a reservoir. Practically all methods make use of equations of the form

$$W = c\psi(P, \theta) \tag{58}$$

where W = cumulative volume of encroached water
P = reservoir pressure
θ = time
ψ = a function of
c = a coefficient, usually of the unit's volume ÷ (pressure × time)

Schilthuis[41] proposed that water encroachment could be calculated by the equation

$$W = \int_{\theta=o}^{\theta=\theta_1} \frac{dW}{d\theta} = c\int_{\theta=o}^{\theta=\theta_1} (P_o - P)d\theta \tag{59}$$

If the reservoir pressure and fluid production rates remain relatively constant during a particular portion of the field's history, the volume of water moving into the reservoir during that period is equal to the volumetric withdrawal of materials. For example, as shown in Table 35, reservoir pressure declined only 5 psi during the six months' period from December 31, 1948, to June 30, 1949. During that period the following volumes were produced and injected into the reservoir:

Produced tank oil and accompanying gas (column 11)	$5.298(10^6)$ bbl
Produced water (column 5)	$1.079(10^6)$ bbl
	$6.377(10^6)$ bbl
Injected gas (column 13)	$-3.929(10^6)$ bbl
Net withdrawal	$2.448(10^6)$ bbl

If the force driving the water into the reservoir is assumed to be equal to the difference in the original reservoir pressure and the existing reservoir pressure, the average value of c is $\dfrac{2.448(10^6)}{181(2750 - 1740)} =$ 13.40 bbl per day per psi.* Using this value of c in connection with a graphical integration of Equation (59) results in the cumulative encroachment curve shown in Figure 69.

The above method developed by Schilthuis involves the assumption that steady state flow conditions exist at all times in the aquifer surrounding

* Note that the fluids in the reservoir expand $0.325(10^6)$ bbl (column 15) while undergoing a pressure reduction from 1740 to 1735 psia. However, this expansion term cannot enter into the evaluation of c because we are presupposing that the quantities of fluids in the reservoir are unknown.

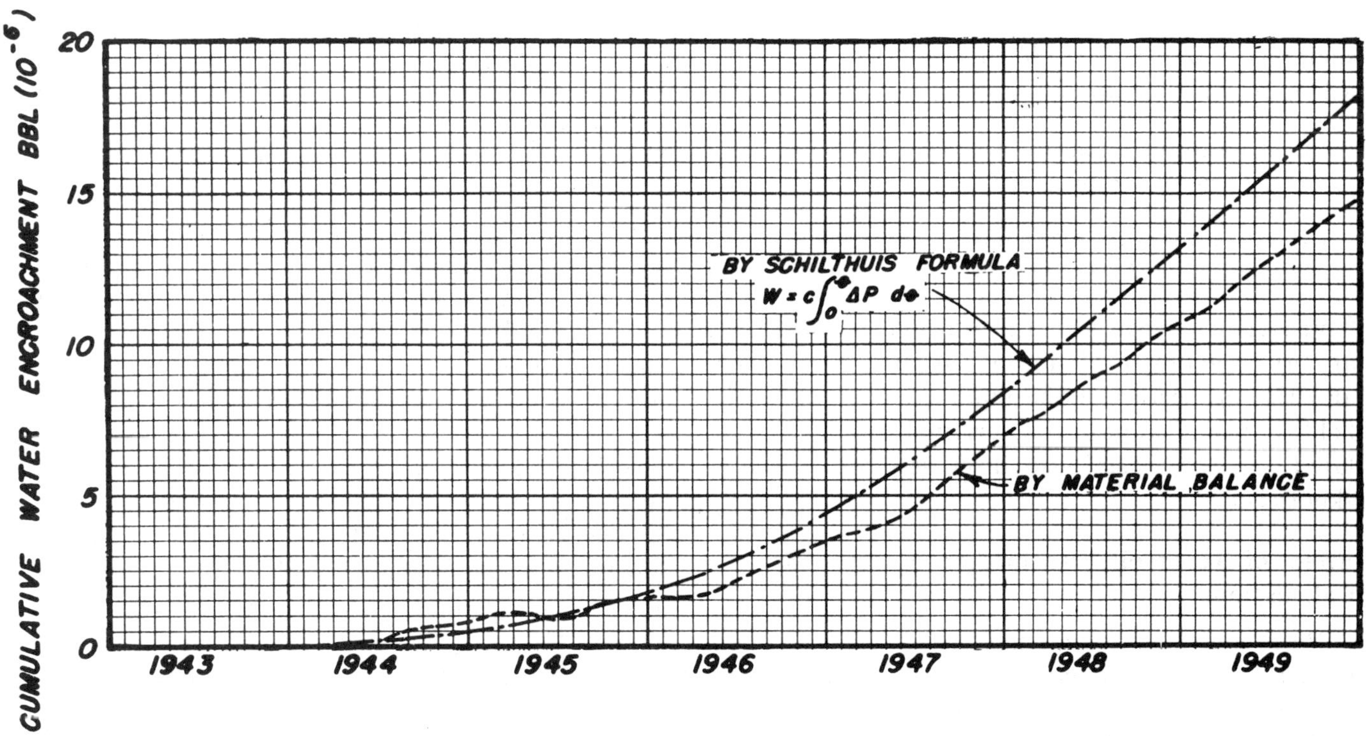

Figure 69—Comparison of Water Encroachment Calculated by Material Balance Method and by Schilthuis Formula

the reservoir. Hurst[46] proposed a method to take into account the transient pressures existing in the reservoir and surrounding aquifer, and developed equations for linear, radial, and spherical flow conditions. In the Hurst method, the value of c is calculated from the permeability and porosity of the rocks containing the aquifer, the compressibility and viscosity of the water, and the cross-sectional area available for water to enter the reservoir.

At first glance it would seem that much could be gained by evaluating the water encroachment independently of the material balance method. To do so would reduce the number of unknowns in Equation (47) to two, and, in many cases where separate evidence indicates that free gas was not originally present in the reservoir, it would allow one to estimate the original oil in place. Old[47] carried out such calculations for the Jones Sand, Schuler Pool, Arkansas. By use of Hurst's equations and assuming different values of c in Equation (58), Old calculated the volume of water that had encroached into the reservoir at different times. For each value of encroached water and time, he could calculate a value for the original oil in place. The particular value of c that gave the minimum deviation (smallest spread) of calculated values of Q_o was accepted as the best factor to use in calculating the water influx term. It so happened that the best value of Q_o was in close agreement with the value determined from geologic and core analysis data.

A somewhat similar approach for calculating the original oil in place has been presented by Brownscombe and Collins.[48] They used the deviation of measured and calculated pressure values for each reserve value in contrast to Old's method of using the deviation of successively calculated values of original oil from the mean value.* To put it another way, Brownscombe and Collins assumed different values of aquifer permeability and original hydrocarbon reserves and determined how the assumed values affected the calculated pressure-time relation of the reservoir. Their work is important because it demonstrates the difficulty in attempting to employ material balance methods and calculated water encroachment quantities in the evaluation of original hydrocarbons in the reservoir. Specifically, a match between the cal-

* Brownscombe and Collins used, in their example, a gas reservoir rather than an oil reservoir. This should help to obtain a better answer as it gets around the question regarding relative proportions of gas and liquid phase originally in the reservoir.

culated pressure curve and the true pressure curve, such that the standard deviation of the difference was only 3 psi, could be obtained by:

(a) An aquifer permeability of 46 millidarcys and an original reserve of Q_o barrels, or

(b) An aquifer permeability of zero and original reserve of 2.6 Q_o barrels.

Thus, when one considers the difficulty of evaluating an "average pressure" during the early life of a field and the properties of the aquifer surrounding the field, it would appear that original reserves calculated on the basis just discussed are open to skepticism.

Effect of Reservoir Structure on Material Balance Calculations

Reservoirs that have thick sands, or reservoirs with thin sands but with steep dips, present their own problems in material balance calculations. In addition to the problem of how best to average the pressures throughout the reservoir, the problem arises of how best to represent the "average" PVT behavior of the hydrocarbon in the reservoir. Not only may the properties of the reservoir fluid differ from place to place, but the fluids can, and often do, vary at different elevations within the reservoir.

Sage and Lacey[49] have presented data from which the change in composition in a static oil column can be calculated as a function of distance below a gas-oil interface. Such differences in fluid composition have been observed in several California fields by noting that the producing gas-oil ratio of wells high on the structure (fluid still undersaturated) is greater than the gas-oil ratio from wells located structurally lower. Also, the °API gravity of the produced tank oil often decreases with depth. Apparently there is no manner by which one can forecast the saturation conditions that exist throughout a reservoir. The method of Sage and Lacey indicates only the limiting case of the saturation-depth behavior, and is based on the existence of gravitational equilibrium between the many hydrocarbon components. For example, calculations for one California reservoir, in which the liquid phase was saturated at 3000 psia and 185°F at the gas-oil interface, indicated that the saturation pressure of the liquid 500 feet below the interface could be as low as 2920 psia. Because of the static pressure exerted by the liquid column, the reservoir pressure was 3160 psia at 500 feet below the gas-oil interface. Thus, the saturation pressure at this depth could vary anywhere between 2920 and 3160 psia.

The importance of correctly weighting PVT data can be appreciated from the previous discussion of material balance methods. For example, a reservoir originally having a liquid undersaturated throughout, but in which the liquid phase is progressively more undersaturated at lower points in the reservoir, should not be represented by a P-V curve that has a distinct "break" at a particular saturation pressure. Insofar as the reservoir behavior is concerned, the liquid elements at different depths go through their bubble points at different times; consequently, any P-V curve weighted to represent the gross reservoir should exhibit a gradual change over the pressure range of all the bubble points.

A second point to be considered in regard to material balance calculations on high relief pools is that of possible gravity drainage effects. Burtchaell[50] points out that the material balance approach involves an assumption that the liquid phase within the sand will not materially change its location during the life of the field. Where steep dips and high permeabilities are involved, so that the liquid and gas phases can migrate toward top and bottom of the reservoir, it becomes necessary to take this factor into account in weighting the PVT properties of the reservoir material.

Summary of Principal Points

The primary object of this chapter was not to discuss in detail the many ways of making material balance calculations. Rather, it was to point out how laboratory PVT data are used in several of the more common methods.

For discussion purposes, material balance equations can be placed into one of two types. The first type, which is exemplified by the work of Tarner, treats the reservoir liquid phase and reservoir gas phase as separate identities. Consequently, one requires the pressure-gas solubility and the pressure-formation volume factor of saturated liquid at pressures over the complete range of the study, as well as the volumetric behavior of the free reservoir gas. The second type is exemplified by the material balance equation published by Schilthuis. Here, the original liquid and the original gas in the reservoir are the two main fluids whose physical properties must be known. Thus, when using the second method, it is necessary to know only the solution gas-oil ratio, formation volume factor, and the pressure-volume relation of the original reservoir liquid phase, and the volumetric behavior of the original reservoir gas phase.

Because the "volumetric balance" must be based in part on the volumes of liquid and gas withdrawn from the reservoir and measured at the surface, the solubility and volumetric factors necessary to convert these surface volumes to "in place" volumes must be obtained from flash tests. Nowhere in the material balance equations exemplified by Equations (54) and (57) do differential data occur directly. True, a picture of the vaporization process occurring in the reservoir may entail differential vaporization, but the calculated volumes are of necessity in terms of flash vaporizations conducted under surface conditions. This requires that formation volumes and solubilities be determined under the trapping conditions that exist in the field. If at some time during the depletion of the reservoir the surface trap conditions are changed, it then becomes necessary to take this into account when evaluating the properties of fluids.

GLOSSARY RELATING TO HYDROCARBON BEHAVIOR

Apparent Convergence Pressure: The pressure at which all the equilibrium constants of a system's components appear to converge to unity when the isothermal log K-log P curves of the components are extrapolated beyond the pressure at which two phases can exist.

Apparent Liquid Density: The apparent ratio of mass to volume exhibited by a dissolved gas as the result of correcting the observed density of a system to 14.7 psia and 60°F and, at that state, subtracting the mass and volume of the liquid components of the system.

Bubble Point: The state of a system characterized by the co-existence of a liquid phase with an infinitesimal quantity of gas phase in equilibrium.

Bubble Point Pressure: The fluid pressure in a system at its bubble point.

Compressibility Factor (**Deviation Factor**): The multiplying factor which must be introduced into the ideal gas law to express the deviation of the gas from ideal behavior ($PV = NZRT$; Z is the compressibility factor).

Condensate (**Distillate**): A liquid resulting from condensation of a vapor phase. In petroleum engineering the term "condensate" is applied specifically to that liquid (crude oil) obtained from hydrocarbon systems existing in the vapor phase in the reservoir.

Cricondentherm: The maximum temperature at which liquid and vapor phases can coexist in equilibrium for a constant composition, multicomponent system.

Critical State: The state of a system at which all properties of the coexisting vapor and liquid phases become identical.

Critical Pressure and/or Temperature: The pressure and/or temperature in a hydrocarbon system at the critical state.

Dew Point: The state of a system characterized by the coexistence of a vapor phase with an infinitesimal quantity of liquid phase in equilibrium.

Dew Point Pressure: The fluid pressure in a system at its dew point.

Differential Process: A process characterized by a constantly changing system composition. A differential process results from the preferential removal from the system of one of the phases, either vapor or liquid, comprising the system.

Differential Gas Liberation: The process whereby the composition of a system is changed by removing gas phase material. By common usage, this term ordinarily refers to the consecutive removal of small quantities of gas phase from a system near its bubble point state.

Dissolved Gas (**Solution Gas**): That material contained in the liquid phase at an elevated pressure and temperature state which ordinarily is gaseous at atmospheric conditions.

Dissolved Gas Drive (**Solution Gas Drive**): A primary recovery process whereby oil is displaced from the reservoir rock by the energy of expansion of the gas originally dissolved in the oil.

Flash Process: A process whereby the quantity of vapor and liquid phases is varied while the overall composition of the system remains constant.

Flash Gas Liberation: The formation of vapor from a liquid under conditions wherein the vapor and liquid phases remain in equilibrium.

Formation Gas-Oil Ratio (**Reservoir Gas-Oil Ratio**): The ratio of the gas to the oil which coexist in the reservoir when they are segregated and reduced to standard atmospheric conditions. This may be restricted to "original" or "current." It ordinarily is expressed as cubic feet per barrel.

Formation-Volume Factor: The volume occupied by one barrel of stock tank oil and its associated gas at an elevated pressure and temperature state (reservoir). Formation volume factors are further specified as to liquid phase (single phase) or gas plus liquid phase (two phase) formation volumes.

Gas Gravity: Ratio of the molecular weight of the gas to the molecular weight of air, 28.97.

Ideal Gas (Perfect Gas): A fictitious gas conforming to the equation of state:

$$P\overline{V} = \frac{RT}{M}$$

in which P is absolute pressure, $\overline{V}$ specific volume, M molecular weight, T absolute temperature, and R the universal gas constant. An ideal gas has a compressibility factor of unity under all conditions (see "Compressibility Factor"). Actual gases never assume all of the properties of an ideal gas, although some gases approximate the ideal behavior at low pressures.

Injection Gas: Gas injected into a reservoir.

Liquid Saturation: The extent to which pores in a reservoir rock are filled with liquid. Saturation ordinarily is expressed as a per cent of pore volume.

Permeability: A measure of the capacity of a porous material to transmit fluid. The unit of permeability is the darcy. A material has the permeability of one darcy when one atmosphere pressure differential across one centimeter length causes a viscous flow of one cubic centimeter per second of a fluid of one centipose viscosity through a cross section of one square centimeter.

Phase: A homogeneous body of material which differs in its intensive properties from its neighboring phases.

Production (Well Effluent): The material which is produced in liquid and/or gaseous phases at the well head.

Production Gas-Oil Ratio: The ratio of natural gas production rate to crude oil production rate, expressed as cubic feet per barrel measured under standard conditions.

Properties, Extensive and Intensive: Properties which are directly proportional to the amount of material making up the system are termed extensive properties. Those which are independent of the quantity of material and therefore describe its condition at a particular state are called intensive properties.

Pseudo-critical Pressure and Temperature: Fictitious critical pressure and temperature values ascribed to a multicomponent system in order that the reduced pressure-volume-temperature states of the system conform to the reduced states of pure gases.

Reduced Pressure and Temperature: The ratio of the pressure in a system to the critical pressure (or pseudo-critical pressure) of the system. Reduced temperature is the ratio of the temperature of a system to the system's critical temperature.

Reservoir Rock: Any rock which contains a commercially exploitable concentration of hydrocarbons.

Residual Oil: The liquid remaining in a PVT cell at the completion of a differential process at or near the reservoir temperature. Also, the liquid remaining in the reservoir at the end of a specified process.

Saturated Liquid: A liquid which is in equilibrium with vapor at a given pressure and temperature state.

Saturated Vapor: A vapor which is in equilibrium with a liquid at a given pressure and temperature state.

Saturation Pressure: The pressure at which vapor and liquid are in equilibrium. Saturation pressure often is used synonymously with bubble point pressure or dew point pressure.

Shrinkage: The decrease in volume of liquid petroleum caused by the release of solution gas and/or by the thermal contraction of the liquid. The shrinkage may be expressed.

(a) as a percentage of the stock tank volume, or
(b) as a percentage of the reservoir oil volume.

Shrinkage Factor: The reciprocal of formation volume factor, expressed as barrels of stock tank oil per barrel of reservoir oil.

Solution Gas-Oil Ratio (Dissolved Gas-Oil Ratio): The volume ratio obtained by dividing the amount of gas which would be liberated from a system existing originally as a liquid by the liquid remaining after the system is reduced to standard atmospheric conditions.

Stock Tank Oil: Crude oil in equilibrium with a portion of its evolved gases at standard atmospheric conditions.

Systems, Homogeneous and Heterogeneous: A body of matter with finite boundaries that represents the material under consideration. In a homogeneous system the intensive properties vary only in a continuous manner with respect to the extent of the system. A heterogeneous system is made up of a number of homogeneous parts, and abrupt changes in the intensive properties occur at the surface of contact of the homogeneous parts of the system.

Undersaturated Fluid: A liquid or vapor capable of holding additional gaseous or liquid components in solution at the specified pressure and temperature.

LIST OF SYMBOLS

A	Arbitrary constant
a	Arbitrary constant
b	Gas constant, per lb
c	Arbitrary constant
F	Total moles of material comprising a system
f	Fugacity (lb per sq in)
G	Standard cu ft of surface gas injected into a reservoir
j	Highest molecular weight component in a system
K	Equilibrium constant, $K = y/x$
k	Permeability
L	Moles of material in a liquid state
M	Molecular weight
m	Weight of material (lb)
m	Ratio of reservoir volume occupied by free gas to reservoir volume occupied by oil
N	Moles of material (lb molecular weights)
n	Mole fraction of a constituent in a system as a whole
$\underline{n}$	Weight fraction of a constituent in a system as a whole
P	Pressure (lb per sq in absolute)
p	Pressure (atmospheres)
p	Vapor pressure of a compound (lb per sq in absolute)
$\underset{\sim}{pp}$	Partial pressure of a constituent in a system (lb per sq in)
Q	Oil volume, bbl at 14.7 psia and 60°F
q	Volumetric rate of flow
R	Universal gas constant, per mole
r	Gas-oil ratio (cu ft per bbl)
S	Gas solubility (cu ft per bbl)
s	Gas solubility (cu ft gas per cu ft liquid)
T	Temperature, degrees Rankine, °R, (°F absolute)
t	Temperature (degrees Fahrenheit, °F)
u	Formation volume factor of gas plus liquid phases (bbl per bbl of tank oil)
u'	Formation volume factor of liquid phase (bbl per bbl of tank oil)
$\bar{u}$	Relative volume factor of gas plus liquid phases (bbl per bbl of residual oil)
$\bar{u}'$	Relative volume factor of liquid phase (bbl per bbl of residual oil)
V	Total volume occupied by a system (cu ft)
$\bar{V}$	Specific volume (cu ft per lb)
$\bar{\mathrm{V}}$	Partial specific volume (cu ft per lb)
$\underset{\sim}{\bar{\bar{\mathrm{V}}}}$	Residual partial specific volume (cu ft per lb)
v	Reservoir gas volume (bbl per standard cu ft)
W	Encroached water (bbl)
w	Produced water (bbl)
x	Mole fraction of a constituent in a liquid phase
y	Mole fraction of a constituent in a gas phase
Z	Compressibility factor
T_c	Critical temperature of a compound, °R
T_R	Reduced temperature, $\frac{T}{T_c}$
pT_R	Pseudo-reduced temperature, $\frac{T}{pT_c}$
P_c	Critical pressure of a compound (psia)
P_R	Reduced pressure, $\frac{P}{P_c}$
pP_R	Pseudo-reduced pressure, $\frac{P}{pP_c}$
log	Common logarithm, base 10
GOR	Gas-oil ratio (cu ft per bbl tank oil)

MCF	One thousand cu ft of gas at 14.7 psia and 60°F.

Greek Symbols

α (alpha)	Coefficient of thermal expansion, volume per volume per °F, $\frac{1}{V}\left(\frac{\delta V}{\delta T}\right)_P$
α (alpha)	Proportional to
β (beta)	Coefficient of compressibility, volume per volume per psi, $\frac{1}{V}\left(\frac{\delta V}{\delta P}\right)_T$
γ (gamma)	Specific gravity, referred to air at 14.7 psia and 60°F
δ (delta)	Specific gravity, referred to water at 14.7 psia and 60°F
θ (theta)	Time (arbitrary units)
μ (mu)	Viscosity (centipoise)
ρ (rho)	Density (lb per cu ft)
ψ (psi)	A function of

Subscripts

$_o$	Original conditions or property
$_1, _2, _3$, etc.	Condition or property at time, $\theta_1, \theta_2, \theta_3$
$_g$	Pertaining to gas phase or gas
$_l$	Pertaining to liquid phase or liquid
$_i$	Injected
$_n$	Net
$_p$	Produced
$_B$	Bubble point condition or property
$_D$	Dew point condition or property

Miscellaneous

5.615	cu ft in a 42 gallon barrel
379	cu ft occupied by one lb mole of a nearly perfect gas at 14.7 psia and 60°F
10.73	Universal gas constant in psia, cu ft, °R, and lb mole units
43,560	cu ft in an acre-foot
7758	bbl of volume in one acre-foot
14.7 psia and 60°F	Standard conditions for measuring gases and liquids

REFERENCES

(1) Kay, W. B.; "Liquid Vapor Phase Equilibrium Relations in the Ethane-n-Heptane System," Ind. Eng. Chem., **30**, 459 (1938)

(2) Katz, D. L. and Kurata, F. U.; "Retrograde Condensation," Ind. Eng. Chem., **32**, 817 (1940)

(3) Bicher, G. B. and Katz, D. L.; "Viscosity of Natural Gases," Trans. AIME, **155**, 246 (1944)

(4) Beal, C.; "The Viscosity of Air, Water, Natural Gas, Crude Oil and Its Associated Gases at Oil Field Temperatures and Pressures," Trans. AIME, **165**, 94 (1946)

(5) Schilthuis, R. J.; "Technique of Securing and Examining Sub-Surface Samples of Oil and Gas," API Drilling and Production Practice, 120 (1935)

(6) Grandstone, P. and Cook, A. B.; "Collecting and Examining Sub-Surface Samples of Petroleum." U. S. Bureau of Mines, T. P. 629 (1941)

(7) Standing, M. B.; "A Pressure-Volume-Temperature Correlation for Mixtures of California Oils and Gases," API Drilling and Production Practice, 275 (1947)

(8) Sage, B. H. and Olds, R. H.; "Volumetric Behavior of Oil and Gas from Several San Joaquin Valley Fields," Trans. AIME, **170**, 156 (1947)

(9) Flaitz, J. M. and Parks, A. S.; "Sampling Gas-Condensate Wells," Trans. AIME **146**, 13 (1942)

(10) Katz, D. L., Brown, G. G., and Parks, A. S.; "NGAA Report on Sampling Two-Phase Gas Streams from High Pressure Condensate Wells" (September, 1945)

(11) van der Waals, J. D.; Thesis, Leiden, 1873 (Dutch) "Continuity of the Gaseous and Liquid State of Matter"

(12) Kay, W. B.; "Density of Hydrocarbon Gases and Vapors at High Temperature and Pressure," Ind. Eng. Chem., **28**, 1014 (1936)

(13) Standing, M. B. and Katz, D. L.; "Density of Natural Gases," Trans. AIME, **146**, 140 (1942)

(14) Mathews, T. A., Roland, C. H., and Katz, D. L.; "High Pressure Gas Measurement," Proceedings NGAA, 41 (1942)

(15) Katz, D. L.; "High Pressure Gas Measurement," Refiner and Natural Gasoline Manufacturer, June, 1942

(16) Eilerts, C. K., Carlson, H. A., and Mullen, N. B.; "Effect of Added Nitrogen on Compressibility of Natural Gas," World Oil, **129**, June, 1948

(17) Olds, R. H., Sage, B. H., and Lacey, W. N.; "Partial Volumetric Behavior of the Methane-Carbon Dioxide System," API Fundamental Research on Occurrence and Recovery of Petroleum, 44 (1943).

(18) Reamer, H. H., Olds, R. H., Sage, B. H., and Lacey, W. N.; "Phase Equilibria in Hydrocarbon Systems—Volumetric Behavior of Ethane-Carbon Dioxide System," Ind. Eng. Chem., **37**, 688 (1945)

(19) Sage, B. H. amd Lacey, W. N.; "Partial Volumetric Behavior of the Lighter Paraffin Hydrocarbons in the Gas Phase," API Drilling and Production Practice, 641 (1939)

(20) Sage, B. H. and Lacey, W. N.; "Volumetric and Phase Behavior of Hydrocarbons," Stanford University Press (1939)

(21) Sage, B. H. and Lacey, W. N.; "Thermodynamic Properties of the Lighter Paraffin Hydrocarbons and Nitrogen," Published by API, 1950

(22) Standing, M. B. and Katz, D. L.; "Density of Crude Oils Saturated with Natural Gas," Trans. AIME, **146**, 159 (1942)

(23) "National Standard Petroleum Oil Tables," National Bureau of Standards Circular C 410 (1936)

(24) NGAA, "Standard Factors for Volumetric Correction and Specific Gravity Conversions of Liquefied Petroleum Gases and Volatile Gasolines," (1942)

(25) Hanson, G. H., Kuist, B. B., and Brown, G. G.; "Liquid Densities of Volatile Hydrocarbon Mixtures," Ind. Eng. Chem., **36**, 1161 (1944)

(26) Sage, B. H., Hicks, B. L., and Lacey, W. N.; "Partial Volumetric Behavior of the Lighter Hydrocarbons in the Liquid Phase," API Drilling and Production Practice, 402 (1938)

(27) Katz, D. L.; "Prediction of the Shrinkage of Crude Oils," API Drilling and Production Practice, 137 (1942)

(28) Muskat, M.; "Physical Principles of Oil Production," McGraw-Hill Book Company, Inc., First Edition, 1949

(29) Souders, M., Selheimer, C. W., and Brown, G. G.; "Equilibria Between Liquid and Vapor Solutions of Paraffin Hydrocarbons," Ind. Eng. Chem., **24**, 517 (1932)

(30) Science of Petroleum, **2**, 1544 (1938) Oxford University Press

(31) Roland, C. H., Smith, D. E., and Kaveler, H. H.; "Equilibrium Constants for a Gas-Distillate System," Oil and Gas Journal, **39**, No 46, 128 (March 7, 1941)

(32) Standing, M. B. and Katz, D. L.; "Vapor Liquid Equilibria of Natural Gas-Crude Oil Systems," Trans. AIME, **155**, 232 (1944)

(33) Katz, D. L. and Hachmuth, K. H., "Vaporization Equilibrium Constants in a Crude Oil-Natural Gas System," Ind. Eng. Chem. **29**, 1072 (1937)

(34) Gilliland, E. R. and Scheeline, H. W.; "High-Pressure Vapor-Liquid Equilibrium for the Systems Propylene-Isobutane and Propane-Hydrogen Sulfide," Ind. Eng. Chem., **32**, 48 (1940)

(35) White, R. R. and Brown, G. G.; "Phase Equilibrium at High Temperatures," Ind. Eng. Chem., **34,** 1162 (1942)

(36) Sage, B. H., Hicks, B. L., and Lacey, W. N.; "Tentative Equilibrium Constants for Light Hydrocarbons," API Drilling and Production Practice, 386 (1938)

(37) Dodson, C. R. and Standing, M. B.; "Prediction of Volumetric and Phase Behavior of Naturally Occurring Hydrocarbon Systems," API Drilling and Production Practice, 326 (1941)

(38) Sage, B. H. and Lacey, W. N.; "Apparatus for Determination of Volumetric Behavior of Fluids," Trans. AIME, **174, 102** (1948)

(39) Rzasa, M. J. and Katz, D. L.; "Calculation of Static Pressure Gradients in Gas Wells," Trans. AIME, **160,** 100 (1945)

(40) Vitter, A. L., Jr; "Back Pressure Tests on Gas-Condensate Wells," API Drilling and Production Practice, 79 (1942)

(41) Schilthuis, R. I.; "Active Oil and Reservoir Energy," Trans. AIME, **118,** 33 (1936)

(42) Tarner, J.; "How Different Size Gas Caps and Pressure Maintenance Programs Affect Amount of Recoverable Oil," Oil Weekly, **114,** No. 2 (1944) 32

(43) Coleman, S., Wilde, H. D., and Moore, T. W.; "Quantitative Effect of Gas-Oil Ratios on Decline of Average Rock Pressure," Trans. AIME, 174 (1930)

(44) Cook, A. B.; Part II, Magnolia Oil Field, Columbia County, Arkansas, "Derivation and Application of Material-Balance Equations," U. S. Bureau of Mines, R. I. 3720 (1943)

(45) Woods, R. W. and Muskat, M.; "An Analysis of Material Balance Calculations," Trans. AIME, **160,** 124 (1945)

(46) Hurst, W.; "Water Influx Into a Reservoir and Its Application to the Equation of Volumetric Balance," Trans. AIME, **151,** 57 (1943)

(47) Old, R. E., Jr.; "Analysis of Reservoir Performance," Trans. AIME, **151,** 26 (1943)

(48) Brownscombe, E. R. and Collins, F.; "Estimation of Reserves and Water Drive from Pressure and Production History," Trans. AIME, **186,** 92 (1949)

(49) Sage, B. H. and Lacey, W. N.; "Gravitational Concentration Gradients in Static Columns of Hydrocarbon Fluids," Trans. AIME, **132,** 120 (1939)

(50) Burtchaell, E. P.; "Reservoir Performance of a High Relief Pool," Trans. AIME, **186,** 171 (1949)

APPENDIX I

GENERAL CORRELATIONS OF HYDROCARBON BEHAVIOR

At several places in the text, use has been made of the generalized PVT correlations reported by the author in 1947.[7] It has been pointed out by others* that the correlating factors were obtained by empirical methods; therefore, it is difficult, if not impossible, to state the accuracy to be expected when the correlations are applied to conditions widely different from those covered by the primary data. This limitation of the empirical method was recognized at the time the original correlations were made. However, in view of the fact that it was desirable to use in the correlations only those values that are readily available from field data, i.e., surface gas gravity, tank oil gravity, gas-oil ratio, pressure, and temperature (reservoir), it was necessary to use the empirical methods.

Of the three correlations presented in Figures 27, 39, and 54, the correlation of bubble point pressure shown in Figure 54 is apt to give the largest errors. Fifty-eight per cent of the 105 experimentally determined bubble point pressures comprising the correlation were within 100 psi of the value given by the correlation, although 12 per cent were farther than 200 psi from the correlation. The arithmetic average errors were 4.8 per cent and 106 psi. The correlation shown in Figure 27 was much more exact; the arithmetic average of the errors was only 1.17 per cent. Ninety-two per cent of the 387 experimental points shown in Figure 37 are within five per cent of the correlation.

To obtain the best use from the empirical correlations, the conditions under which they are used should be close to the conditions on which the correlations are based. These conditions can be summarized as follows:

(1) Gas-oil ratios are results of flash vaporizations at surface conditions. The ratio is expressed as total cubic feet of surface gas per barrel of tank oil.

(2) Gas gravity used in the correlation is the gravity of all gases produced at the surface, including tank vapors.

(3) The surface flash process corresponds to a two-stage separation process at, or near, 100°F, the first stage ordinarily being within the pressure range of 150 to 450 psi, and the second at atmospheric pressure.

(4) The systems comprising the correlations did not contain appreciable quantities of nitrogen or carbon dioxide. The bubble point pressure of systems containing these compounds ordinarily will be higher than indicated by the correlation in Figure 54.

(5) Formation volume data are in terms of a unit barrel of tank oil, measured at 14.7 psia and 60°F.

The correlating functions in Figures 27, 39, and 54 are not easy to handle. The function shown in Figure 39 is particularly difficult. Therefore, the correlations have been transformed into the three calculating charts found in the pocket on the back cover. Values read from these charts have the same accuracy as values calculated from the original correlations. Chart 1 presents the total formation volume of systems at various pressure and temperature states. This chart is applicable to dissolved gas systems at pressures below their bubble point pressures and to gas condensate systems at pressures both below and above their dew point pres-

* See comments by B. H. Sage and W. Tempelaar Lietz, pp. 279–287. 1947 API Drilling and Production Practice.

sures. Chart 2 presents the bubble point pressure of a dissolved gas system as a function of gas-oil ratio, gas gravity, tank oil gravity, and temperature. The chart also may be used in the reverse direction to estimate the solution gas-oil ratio of a system from its saturation pressure and temperature. Chart 3 allows one to calculate the formation volume of bubble point liquids. From these three calculating charts, it is possible to make engineering estimates of the physical properties of multi-component systems under conditions existing in petroleum reservoirs.

APPENDIX II

CHART EQUATIONS

At the time this book was prepared (1950), two important tools of the petroleum engineer were a 10-inch slide rule, and a compilation of charts, tables, and miscellaneous information. In fact, the large size of the book was selected specifically to permit the use of large-size charts.

Today, slide rules have pretty much been replaced by small electronic calculators and by large computers. This reduces the need for charts but opens up the need for mathematical expressions of the charts. The reprinting of the book by the Society of Petroleum Engineers of AIME offered the opportunity to "update" it by adding this Appendix.

Most of the equations that follow were developed by the author by simple curve-fit procedures. They are not the most precise equations that can be developed but are considered to be adequate for their intended use. The goodness of fit of the equations to the originating chart is indicated by the average percent difference of samples of chart and equation values and the standard deviation of the percent differences. Note that these criteria apply to values read from charts and do not indicate the degree of agreement between original data values and equation values.

The format of the Appendix is simple. Equations will be given in the order of the corresponding chart's appearance in the book — the gas compressibility factor chart on Page 24 being the first. In addition, equations are given at the end of the Appendix with which to calculate viscosity of reservoir oils and gases.

Figure 15, Page 24. Compressibility Factors for Natural Gases

The Standing-Katz[1] compressibility factor chart has been set up for computer solution by a number of individuals. Takacs[2] recently published a comparison of eight computer programs to calculate Z-factors as a function of pseudoreduced pressure and temperature.[3-11] His comparison of 180 chart and calculated values showed that the Dranchuk and Abou-Kassam equations[11] had the smallest absolute error, 0.316 percent. A FORTRAN program for this calculation is given in the original article and will not be repeated here.

A modification of the equations published by Brill and Beggs[9] yield compressibility factor values accurate enough for many engineering calculations. They have the additional advantage of being programmable for the HP-67 and TI-SR52 calculators. The equations are:

$$Z = A + (1-A)/e^{B} + C\,p_{pr}^{\,D} \quad (1)$$

where

$$A = 1.39\,(T_{pr}-0.92)^{0.5} - 0.36\,T_{pr} - 0.101 \quad (2)$$

$$B = (0.62-0.23\,T_{pr})p_{pr} + \left[\frac{0.066}{(T_{pr}-0.86)} - 0.037\right] p_{pr}^{2} + \frac{0.32}{10^{9(T_{pr}-1)}}\,p_{pr}^{6} \quad (3)$$

$$C = (0.132 - 0.32\log T_{pr}) \quad (4)$$

$$D = \text{antilog}\,(0.3106 - 0.49\,T_{pr} + 0.1824\,T_{pr}^{2}) \quad (5)$$

Agreement between calculated Z values and chart Z values was tested at T_{pr} values of 1.2, 1.3, 1.4, 1.5, 1.7, 2.0, and 2.4 and p_{pr} values of 0, 1, 2, 3, 4,

5, 7, 9, 11, and 13. Results were as follows.

	T_{pr}, p_{pr} Range	
	$1.3 < T_{pr} < 1.7$ $0 < p_{pr} < 5$	$1.2 < T_{pr} < 2.4$ $0 < p_{pr} < 13$
Number of points	24	70
Δ, percent	0.19	0.02
$S(\Delta)$, percent	0.96	1.18
Δ_{max}, percent	+2.63	−3.93

The above relationships are not valid outside the reduced temperature range 1.2 to 2.4.

Figure 17, Page 26. Pseudocritical Constants of Gases and Condensate Fluids

The pseudocritical pressure and temperature-gas gravity curves of Figure 17 represent mixtures having negligible amounts of N_2, CO_2, and H_2S. The "California Gases" curves came from Figure 16 and represent surface trap gases and stock-tank vapors whose high gravity results from relatively high proportions of ethane, propane, and butanes. The "condensate fluids"[1] were laboratory generated gases in equilibrium with crude oil at high pressures (1,000 to 8,000 psi). Their high gravities come from relatively large quantities of heptanes and heavier compounds. In general, it seems most reasonable to use the "condensate" curves in calculations that involve gases in equilibrium with crude oil of condensate in petroleum reservoirs and to use the "California" curves for surface gases. However, it should be kept in mind that these curves represent only four of an infinite number of possible curves. There is no unique correlation of pseudocritical constants with gas gravity.

The surface, or "California" gas curves are fit by the equations:

$$p_{pcHC} = 677 + 15.0\,\gamma_{gHC} - 37.5\,\gamma_{gHC}^2 \tag{6}$$

$$T_{pcHC} = 168 + 325\,\gamma_{gHC} - 12.5\,\gamma_{gHC}^2 \tag{7}$$

The reservoir, or "condensate" gas curves, has the following relationship:

$$p_{pcHC} = 706 - 51.7\,\gamma_{gHC} - 11.1\,\gamma_{gHC}^2 \tag{8}$$

$$T_{pcHC} = 187 + 330\,\gamma_{gHC} - 71.5\,\gamma_{gHC}^2 \tag{9}$$

The pseudocritical values of gases that contain N_2, CO_2, and H_2S can be calculated by the method of Wichert and Aziz.[12] If γ_{gM} is the gravity of the whole gas mixture, the gravity of the hydrocarbon portion is

$$\gamma_{gHC} = \frac{\gamma_{gM} - 0.967\,y_{N_2} - 1.52\,y_{CO_2} - 1.18\,y_{H_2S}}{1 - y_{N_2} - y_{CO_2} - y_{H_2S}} \tag{10}$$

The pseudocriticals of the whole gas mixture are then

$$p_{pcM} = (1 - y_{N_2} - y_{CO_2} - y_{H_2S})\,p_{pcHC} + 493\,y_{N_2} + 1{,}071\,y_{CO_2} + 1{,}306\,y_{H_2S} \tag{11}$$

$$T_{pcM} = (1 - y_{N_2} - y_{CO_2} - y_{H_2S})\,T_{pcHC} + 227\,y_{N_2} + 548\,y_{CO_2} + 672\,y_{H_2S} \tag{12}$$

These mixture pseudocriticals should be adjusted for their CO_2 and H_2S content by means of the Wichert-Aziz equations:

$$\epsilon = 120\left[(y_{CO_2} + y_{H_2S})^{0.9} - (y_{CO_2} + y_{H_2S})^{1.6}\right] + 15\,(y_{CO_2}{}^{0.5} - y_{H_2S}{}^{4}) \tag{13}$$

$$T'_{pcM} = T_{pcM} - \epsilon \tag{14}$$

$$p'_{pcM} = \frac{p_{pcM}\,(T_{pcM} - \epsilon)}{T_{pcM} + y_{H_2S}\,(1 - y_{H_2S})\,\epsilon} \tag{15}$$

Note that in Equation (10) $\gamma_{gHC} \geqslant 0.55$ (methane gravity). In Equations (11 and 12) the pseudocriticals of the hydrocarbon portion, p_{pcHC} and T_{pcHC}, may be represented by Equations (6 and 7 or 8 and 9), depending on the type of hydrocarbon gas involved.

Figure 23, Page 35. Pseudo-Liquid Density of Systems Containing Methane and Ethane

Figure 23 makes use of the apparent liquid density-system density relationships shown in Figure 22.[13] The relationship between the propane plus density, ρ_{3+}, the ethane plus density, ρ_{2+}, and the weight percent ethane in the ethane plus portion of the system, w_2, is

$$\rho_{3+} = \frac{(1 - 0.01\,w_2)}{\dfrac{1}{\rho_{2+}} - \dfrac{0.01\,w_2}{15.3 + 0.3167\,\rho_{2+}}} \tag{16}$$

Similarly, the relationship between the ethane plus density, ρ_{2+}, the methane plus density, ρ_{1+} (total system pseudo-liquid density), and the weight percent methane in the whole system, w_1, is

$$\rho_{2+} = \frac{(1 - 0.01\,w_1)}{\dfrac{1}{\rho_{1+}} - \dfrac{0.01\,w_1}{0.312 + 0.450\,\rho_{1+}}} \tag{17}$$

Naturally occurring reservoir and surface oils do not have large weight percents of dissolved ethane and methane. The values of $w_2 = 1.05$; $w_1 = 5.4$ shown in Table 7, Page 38, are typical of a reservoir oil having a bubble-point pressure of about 3,500 psia. Thus, there is limited use for a large portion of Figure 23. Furthermore, the form of Equations (16 and 17) is not the best for solution for ρ_{2+} and ρ_{1+}. Therefore, values obtained from Equations (16 and 17) were curve fit in the region $40<\rho_{3+}<60$; $0<w_2<10$; $0<w_1<16$, with the following results:

$$\rho_{2+} = \rho_{3+}(1-0.01386\,w_2-0.000082\,w_2^2) + 0.379\,w_2 + 0.0042\,w_2^2 \quad (18)$$

$$\rho_{1+} = \rho_{2+}(1-0.012\,w_1-0.000158\,w_1^2) + 0.0133\,w_1 + 0.00058\,w_1^2 \quad (19)$$

Densities calculated from Equations (18 and 19) agree with densities calculated from Equations (16 and 17) within 0.05 lb/cu ft and are more precise than values read from Figure 23.

Figure 24, Page 36. Density Correction for Compressibility of Liquids

The units of density (lb/cu ft) to be added to the 14.7 psia/60°F pseudo-liquid density (Figure 23) to obtain a pseudo-liquid density at pressure p and 60°F is given by the relationship:

$$\Delta\rho_T = \left(0.167 + 16.181\cdot 10^{-0.0425\,\rho_{pL}}\right)\left(\frac{p}{1{,}000}\right) - 0.01\left(0.299 + 263\cdot 10^{-0.0603\,\rho_{pL}}\right)\left(\frac{p}{1{,}000}\right)^2 \quad (20)$$

where ρ_{pL} = pseudo-liquid density at atmospheric conditions.

The following tabulation shows the agreement between $\Delta\rho_T$ values calculated from Equation (20) and values read from Figure 24 at pressures of 2,000, 4,000, 6,000, 8,000, and 10,000 psia.

ρ_{pL}	$\bar{\Delta}$ (percent)	$S(\Delta)$ (percent)	ρ_{nl}	$\bar{\Delta}$ (percent)	$S(\Delta)$ (percent)	
60	−0.09	1.40	35	−2.63	1.22	
55	0.57	0.83	30	−4.94	2.31	
50	0.49	1.20	25	−6.89	4.89	
45	0.29	1.07	40-60	0.05	1.21	(25 values)
40	−0.90	1.23	25-60	−1.77	3.29	(40 values)

Figure 25, Page 37. Density Correction for Thermal Expansion of Liquids

The units of density (lb/cu ft) to be subtracted from the pseudo-liquid density value at pressure p and 60°F when the liquid is heated to temperature T is given by the equation:

$$\Delta\rho_p = \left[0.0133 + 152.4\,(\rho_{pL}+\Delta\rho_T)^{-2.45}\right](T-60) - \left[8.1(10^{-6}) - 0.0622\cdot 10^{-0.0764(\rho_{pL}+\Delta\rho_T)}\right](T-60)^2 \quad (21)$$

where $(\rho_{pL}+\Delta\rho_T)$ = pseudo-liquid density at pressure p and 60°F.

The agreement between calculated values of $\Delta\rho_p$ and those read from Figure 25 at temperatures of 80°, 100°, ... , 240°F is shown in the following tabulation:

$(\rho_{pL}+\Delta\rho_T)$	$\bar{\Delta}$ (percent)	$S(\Delta)$ (percent)	
60	0.58	2.04	
55	−0.68	0.75	
50	−0.65	1.53	
45	−0.27	3.42	
40	−1.45	1.74	
40 to 60	−0.50	2.10	(50 values)

Figure 26, Page 40. Apparent Liquid Densities of Natural Gases

The chart by Katz[14] showing the apparent liquid density of a natural gas of gravity γ_g when dissolved in oil of API degrees gravity has the relationship:

$$\rho_{aL} = 38.52\cdot 10^{-0.00326\,°\text{API}} + (94.75-33.93\log °\text{API})\log\gamma_g \quad (22)$$

The following tabulation shows the agreement between chart and equation values at values of gas gravity of 0.6, 0.7, ... , 1.4.

API	$\bar{\Delta}$ (percent)	$S(\Delta)$ (percent)	
20	−0.61	1.05	
30	0.17	0.61	
40	0.64	0.41	
50	0.79	0.21	
60	−0.43	1.13	
20 to 60	0.11	0.93	(45 values)

Figure 27, Page 42. Formation Volume Factor of Bubble-Point Liquids From Gas-Oil Ratio, Dissolved Gas Gravity, Tank Oil Gravity, and Temperature

Figure 27 shows the correlation of bubble-point liquid formation volume factor, B_{ob}, and a correlating number composed of dissolved gas-oil ratio, dissolved gas gravity, tank oil specific gravity, and temperature. This correlation is the basis for Chart 3 in the back of the book. The relationship:

$$B_{ob} = 0.9759 + 12\,(10^{-5})\,(CN)_{B_{ob}}{}^{1.2} \tag{23}$$

where

$$(CN)_{B_{ob}} = R_s\left(\frac{\gamma_g}{\gamma_o}\right)^{0.5} + 1.25\,T \tag{24}$$

fits the curve of Figure 27 with an average difference of 0.01 percent and a standard deviation of differences of 0.34 percent. (Eight point in the range 150 $< (CN) <$ 1,880.)

Figure 37, Page 60. Pseudocritical Temperatures and Pressures for Heptanes and Heavier

The correlation of pseudocritical temperature and pressure of C_{7+} as functions of molecular weight, M, and specific gravity, γ_o, reported by Mathews, Roland, and Katz[15] can be fit by the following expressions:

$$T_{pcC_{7+}} = 608 + 364\log(M-71.2) + (2{,}450\log M - 3{,}800)\log\gamma_o \tag{25}$$

$$p_{pcC_{7+}} = 1{,}188 - 431\log(M-61.1) + \left[2{,}319 - 852\log(M-53.7)\right](\gamma_o - 0.8) \tag{26}$$

The goodness of fits are indicated by the following tabulation:

	$T_{pcC_{7+}}$	$p_{pcC_{7+}}$
Number of Values	28	25
$\bar{\Delta}$	−0.29%	−0.51%
$S(\Delta)$	0.97%	1.61%

Figure 38, Page 61. Effect of Condensate Volume on the Ratio of Surface Gas Gravity to Well Fluid Gravity

Figure 38 is used to obtain the gravity of a reservoir gas condensate system knowing the gravities of the gas and condensate produced at the surface and their producing gas-oil ratio. The chart results from a balance of moles and pounds of surface products but involves the correlation of condensate molecular weight and °API shown in the small insert graph. The equation that represents this balance is

$$\frac{\gamma_w}{\gamma_g} = \left(\frac{240-2.22°\text{API}}{28.97\,\gamma_g}\right)\left[\frac{28.97\,\gamma_g\,(131.5+°\text{API})+18.76\,B}{31{,}560-52.2°\text{API}-2.22\,°\text{API}^2+18.76\,B}\right] \tag{27}$$

Figure 35 is a chart representation of this equation.

Eilerts[16] has developed a relationship that does the same thing. Eilerts' equation makes use of the "equivalent vapor volume" of the produced condensate rather than molecular weight. A curve fit of his equivalent vapor volume vs °API data yields the following equation.

$$\frac{\gamma_w}{\gamma_g} = \frac{\gamma_g + \dfrac{0.6505\,B}{(131.5+°\text{API})}}{\gamma_g\left[1 + B\cdot 10^{-6}(251+9.75\,°\text{API}-0.006\,°\text{API}^2)\right]} \tag{28}$$

Following is a comparison of well fluid gravities calculated from Equations (27 and 28). It will be seen that the two equations yield essentially the same values.

			Calculated Well Fluid Gravity, γ_w	
γ_g	B	°API	Eq. (27)	Eq. (28)
0.6	10	40	0.6332	0.6339
0.6	50	40	0.7613	0.7655
0.6	100	40	0.9117	0.9211
0.8	10	60	0.8263	0.8272
0.8	50	60	0.9266	0.9319
0.8	100	60	1.043	1.042

Figure 54, Page 86. Correlation of Bubble-Point Pressure With Dissolved Gas-Oil Ratio, Dissolved Gas Gravity, Tank Oil Gravity, and Temperature

The bubble-point pressure/bubble-point pressure correlating number curve shown in Figure 54 has the relationship:

$$p_b = 18.2\left[(CN)_{p_b} - 1.4\right] \tag{29}$$

where

$$(CN)_{p_b} = \left(\frac{R_s}{\gamma_g}\right)^{0.83}\cdot 10^{(0.00091\,T - 0.0125°\text{API})} \tag{30}$$

This relationship is the basis for the bubble-point pressure chart discussed in Appendix I and located in the back of the book.

Reservoir Oil Viscosity Correlations of Beal and of Chew and Connally

Estimating reservoir oil viscosity has been done primarily by means of the correlations of Beal[17] and Chew and Connally[18] until recently. Beal's correlation is used to obtain the gas-free crude oil viscosity (stock-tank oil or ''dead oil'') at reservoir temperature as a function of its API gravity. This ''dead oil'' is then adjusted for the amount of dissolved gas it contains at reservoir pressure by means of Chew and Connally's correlation. Should the reservoir pressure be greater than the bubble-point pressure of the hydrocarbon system, a further adjustment is made to the bubble-point liquid viscosity to account for the degree of undersaturation in the reservoir.

The more recent work of Beggs and Robinson[19] apparently gives a better estimate of saturated reservoir oil viscosity. However, it should be pointed out that neither Beal's nor Beggs and Robinson's correlations take into account the chemical nature of the hydrocarbons that make up the crude oil part of the reservoir oil. Chemical nature is an important parameter in correlating liquid hydrocarbon viscosity behavior, and is usually indicated by the UOP Characterization Factor. Until some factor that expresses chemical nature of the crude oil is included in reservoir oil viscosity correlations, it is doubtful that much improvement will be made in developing more precise correlations.

Beal's correlation of gas-free crude oil (dead oil) viscosity as a function of temperature and API gravity is not particularly precise. The average difference between chart and measured values (his Figure 8) amounted to 24 percent. However, accepting his Figure 8 as drawn, the following equations represent a reasonable fit.

$$\mu_{oD} = \left(0.32 + \frac{1.8(10^7)}{°\text{API}^{4.53}}\right)\left(\frac{360}{T+200}\right)^a \tag{31}$$

where

$$a = \text{antilog}\left(0.43 + \frac{8.33}{°\text{API}}\right) \tag{32}$$

The fit of the above equations and Beal's Figure 8 is indicated by the average difference (25 values at °API of 15, 25, ... , 55) of −1.58 percent and a standard deviation of differences of 9.4 percent. The large standard deviation results primarily from the 15° API values. If these are left out, the equations fit the remaining 20 values with $\bar{\Delta} = 1.62$ percent; $S(\Delta) = 4.92$ percent.

Chew and Connally's equation for the viscosity of saturated oil (bubble-point oil) is

$$\mu_{ob} = A\,(\mu_{oD})^b \tag{33}$$

where A and b are functions of solution gas-oil ratio. Fit of A and b values given in their Table 3 are

$$A = \text{antilog}\left\{R_s\left[2.2\,(10^{-7})\,R_s - 7.4\,(10^{-4})\right]\right\} \tag{34}$$

$$b = \frac{0.68}{10^{8.62(10^{-5})R_s}} + \frac{0.25}{10^{1.1\,(10^{-3})R_s}} + \frac{0.062}{10^{3.74\,(10^{-3})R_s}} \tag{35}$$

Equations (34 and 35) reproduce the 12 A and b values in Table 3 with the following precision:

Coefficient	$\bar{\Delta}$ (percent)	$S(\Delta)$ (percent)
A	−0.21	2.79
b	−0.10	0.36

Beal's Figure 11 shows a correlation of the slope of the viscosity-pressure curves in the undersaturated oil region as a function of bubble-point viscosity. A fit of his curve is

$$\frac{\mu_o - \mu_{ob}}{0.001\,(p - p_b)} = 0.024\,\mu_{ob}^{1.6} + 0.038\,\mu_{ob}^{0.56} \tag{36}$$

For 11 values used in developing this equation, $\bar{\Delta} = 0.01$ percent, $S(\Delta) = 4.64$ percent.

Reservoir Gas Viscosity Correlation of Carr, Kobayashi, and Burrows

Reservoir gas viscosity is usually estimated by means of the correlations of Carr, Kobayashi, and Burrows.[20] Their correlation involves a two-step procedure: the gas viscosity at reservoir temperature and atmospheric pressure is estimated first from its gravity and nonhydrocarbon component content. The atmospheric value is then adjusted to reservoir conditions by means of a correction factor based on the reduced temperature and pressure state of the gas in the reservoir.

The atmospheric pressure viscosity, μ_1, at gas gravity values less than 1.5 (region of interest in reservoir engineering calculations) as read from their Figure 6 can be expressed as:

$$\mu_1 = (\mu_1 \text{ uncorrected}) + (N_2 \text{ correction}) + (CO_2 \text{ correction}) + (H_2S \text{ correction}) \quad (37)$$

where

$$(\mu_1 \text{ uncorrected}) = \left[1.709(10^{-5}) - 2.062(10^{-6})\,\gamma_g\right] T + 8.188(10^{-3}) - 6.15(10^{-3}) \log \gamma_g \quad (38)$$

$$(N_2 \text{ correction}) = y_{N_2}\left[8.48(10^{-3}) \log \gamma_g + 9.59(10^{-3})\right] \quad (39)$$

$$(CO_2 \text{ correction}) = y_{CO_2}\left[9.08(10^{-3}) \log \gamma_g + 6.24(10^{-3})\right] \quad (40)$$

$$(H_2S \text{ correction}) = y_{H_2S}\left[8.49(10^{-3}) \log \gamma_g + 3.73(10^{-3})\right] \quad (41)$$

The fit between Equation (38) and Carr *et al.*'s Figure 6 is indicated by $\bar{\Delta} = 0.38$ percent, $S(\Delta) = 0.46$ percent for 30 values in the ranges $0.55 < \gamma_g < 1.55$; $100 < T < 300°F$.

Dempsey[21] has developed computer programs for both μ_1 and the ratio μ_g/μ_1. His relationship for the ratio term is

$$\ln\left(\frac{\mu_g}{\mu_1} \cdot T_{pr}\right) = a_0 + a_1 p_{pr} + a_2 p_{pr}^2 + a_3 p_{pr}^3 + T_{pr}(a_4 + a_5 p_{pr} + a_6 p_{pr}^2 + a_7 p_{pr}^3) + T_{pr}^2(a_8 + a_9 p_{pr} + a_{10} p_{pr}^2 + a_{11} p_{pr}^3) + T_{pr}^3(a_{12} + a_{13} p_{pr} + a_{14} p_{pr}^2 + a_{15} p_{pr}^3) \quad (42)$$

where

$a_0 =$	−2.462 118 20E −00	$a_8 =$	−7.933 856 84E −01
$a_1 =$	2.970 547 14E −00	$a_9 =$	1.396 433 06E −00
$a_2 =$	−2.862 640 54E −01	$a_{10} =$	−1.491 449 25E −01
$a_3 =$	8.054 205 22E −03	$a_{11} =$	4.410 155 12E −03
$a_4 =$	2.808 609 49E −00	$a_{12} =$	8.393 871 78E −02
$a_5 =$	−3.498 033 05E −00	$a_{13} =$	−1.864 088 48E −01
$a_6 =$	3.603 730 20E −01	$a_{14} =$	2.033 678 81E −02
$a_7 =$	−1.044 324 13E −02	$a_{15} =$	−6.095 792 63E −04
		(=	0.00 060 957 9263)

Dempsey's relationship expressed by Equation (42) can be scaled to operate successfully on Texas Instruments' SR-52 and Hewlett-Packard's HP-65 and HP-67 programmable calculators by limiting the above coefficients to four significant figures.

Nomenclature

A = variable in Equations (33 and 34)
a = variable in Equations (31 and 32)
$a_0 \ldots a_{15}$ = coefficients in Equation (42)
B = condensate — gas ratio, bbl/MMcf
B_{ob} = oil formation volume factor at bubble-point state, bbl/bbl STO
b = variable in Equations (33 and 35)
$(CN)_{B_{ob}}$ = bubble-point formation volume factor correlating number
$(CN)_{pb}$ = bubble-point pressure correlating number
ϵ = correction term in Wichert-Aziz Equations (13, 14, and 15), °R
γ_g = gas gravity, air = 1
γ_o = oil specific gravity, water = 1
γ_w = well production gravity, air = 1
log = logarithm base 10
M = molecular weight of heptanes and heavier material, lb/lb mole
p = pressure, psia
p_{pc} = pseudocritical pressure, psia
p'_{pc} = adjusted pseudocritical pressure, psia
p_{pr} = pseudoreduced pressure, dimensionless
R_s = solution (dissolved) gas-oil ratio, cu ft/bbl
ρ_{pL} = pseudo-liquid density at 14.7 psia and 60°F, lb/cu ft
ρ_{aL} = apparent liquid density of dissolved gas at 14.7 psia and 60°F, lb/cu ft
ρ_{1+} = density of methane plus fraction, lb/cu ft
ρ_{2+} = density of ethane plus fraction, lb/cu ft
ρ_{3+} = density of propane plus fraction, lb/cu ft
$\Delta\rho_T$ = change in pseudo-liquid density resulting from compression at 60°F, lb/cu ft
$\Delta\rho_p$ = change in pseudo-liquid density resulting from heating at elevated pressure, lb/cu ft
T = temperature, °F
T_{pc} = pseudocritical temperature, °R
T'_{pc} = adjusted pseudocritical temperature, °R
T_{pr} = pseudoreduced temperature, dimensionless
μ_1 = gas mixture viscosity at 14.7 psia and reservoir temperature, cp
μ_g = gas viscosity at reservoir temperature and pressure, cp
μ_o = oil viscosity at reservoir temperature and pressure, cp
μ_{ob} = oil viscosity at bubble-point state, cp
μ_{oD} = viscosity of gas-free crude oil (dead oil), cp
w_1 = weight percent methane in liquid mixture
w_2 = weight percent ethane in ethane plus fraction of liquid mixture
y = mole fraction of component in gas phase
Δ = percent difference = [(calculated value − chart value)/chart value] · 100
$\bar{\Delta}$ = average percent difference = $1/n\,\Sigma\,\Delta$
$S(\Delta)$ = standard deviation of differences $= \sqrt{(1/n-1)\,(\Sigma\Delta^2) - 1/n\,(\Sigma\Delta)^2}$

Subscripts

M = mixture
HC = hydrocarbon portion of mixture
L = liquid

References

(1) Standing, M. B. and Katz, D. L.: "Density of Natural Gases," *Trans.*, AIME (1942) **146,** 140-149.

(2) Takacs, G.: "Comparisons Made for Computer Z-Factor Calculations," *Oil and Gas J.* (Dec. 20, 1976) 62.

(3) Grey, E. H. and Sims, H. L.: "Z-Factor Determinations in a Digital Computer," *Oil and Gas J.* (July 20, 1959) 80.

(4) Sarem, A. M.: "Z-Factor Equation Developed for Use in Digital Computers," *Oil and Gas J.* (Sept. 18, 1961) 118.

(5) Carlile, R. E. and Gillett, B. E.: "Digital Solutions of an Integral," *Oil and Gas J.* (July 19, 1971) 68.

(6) Papay, J.: "A Termeléstechnológiai Paraméterek Változása a Gástelepek Müvelése Során," *OGIL Müsz, Tud, Küzl.*, Budapest (1968) 267.

(7) Hall, K. R. and Yarborough, L.: "A New Equation of State for Z-Factor Calculations," *Oil and Gas J.* (June 18, 1973) 82.

(8) Yarborough, L. and Hall, K. R.: "How to Solve Equation of State for Z-Factors," *Oil and Gas J.* (Feb. 18, 1974) 86.

(9) Brill, J. P. and Beggs, H. D., University of Tulsa: "Two-Phase Flow in Pipes," INTERCOMP Course, The Hague, 1974.

(10) Dranchuk, P. M., Purvis, R. A., and Robinson, D. B.: "Computer Calculations of Natural Gas Compressibility Factors Using the Standing and Katz Correlation," Institute of Petroleum Technical Series, No. IP 74-008 (1974).

(11) Dranchuk, P. M. and Abou-Kassem, J. H.: "Calculation of Z-Factors for Natural Gases Using Equations of State," *J. Cdn. Pet. Tech.* (July-Sept. 1975) 34.

(12) Wichert, E. and Aziz, K.: "Calculate *Z's* for Sour Gases," *Hydrocarbon Processing* (May 1972) **51,** 119.

(13) Standing, M. B. and Katz, D. L.: "Density of Crude Oils Saturated with Natural Gas," *Trans.*, AIME (1942) **146,** 159-165.

(14) Katz, D. L.: "Prediction of the Shrinkage of Crude Oils," *Drill. and Prod. Prac.*, API (1942) 137-147.

(15) Mathews, T. A., Roland, C. H., and Katz, D. L.: "High Pressure Gas Measurement," *Proc.*, NGAA (1942) **41**.

(16) Eilerts, C. K.: "Phase Relations of Gas Condensate Fluids," *Monograph 10,* USBM (1957).

(17) Beal, C.: "The Viscosity of Air, Water, Natural Gas, Crude Oil and Its Associated Gases at Oil-field Temperatures and Pressures," *Trans.*, AIME (1946) **165,** 94-115.

(18) Chew, J. N. and Connally, C. A., Jr.: "A Viscosity Correlation for Gas-Saturated Crude Oils," *Trans.*, AIME (1959) **216,** 23-25.

(19) Beggs, H. D. and Robinson, J. R.: "Estimating the Viscosity of Crude Oil Systems," *J. Pet. Tech.* (Sept. 1975) 1140-1141.

(20) Carr, N. L., Kobayashi, R., and Burrows, D. B.: "Viscosity of Hydrocarbon Gases under Pressure," *Trans.*, AIME (1954) **201,** 264-272.

(21) Dempsey, J. R.: "Computer Routine Treats Gas Viscosity as a Variable," *Oil and Gas J.* (Aug. 16, 1965) 141.

INDEX